ARMAND MILLET and HIS VIOLETS

ARMAND MILLET AND HIS VIOLETS

Including

VIOLETTES,
leurs origines, leurs cultures

by Armand Millet

Translated and annotated
by

E.J.PERFECT

Park Farm Press

© 1996 E.J.Perfect
Park Farm Press, Glenwood, Park Farm Road,
High Wycombe, HP12 4AF

ISBN 0 9527998 0 4

Cover picture: The violet 'La France' from *Les Violettes* and Armand Millet's signature.
Cover map: Bourg-la-Reine showing the Millet nursery.

Printed in Great Britain
by Whitstable Litho Printers Ltd., Millstrood Road,
Whitstable, Kent CT5 3PP

CONTENTS

	page
Illustrations	vii
Armand Millet	1
The Nursery	25
Exhibiting Violets	26
Violets, their origins and cultivation by Armand Millet. Translated by E.J.Perfect	30
Translator's notes	146
Violets raised or introduced by Millet, his father and his son Lionnel	175
Violets listed by Lionnel Millet in his catalogue for 1932 to 1933	177
Tree Violets	180
Appendix 1 *Tree Violets* by D.Beaton (*The Cottage Gardener* 1848)	189
Appendix 2 *The Tree Violet* by Alpha (*The Gardener's Chronicle*, 1852)	190
Appendix 3 'Mme Arène' grown on a stem (Tree Violet) Millet (*JSNHF* 1902)	192
Appendix 4 A Note on Violets grown as "Trees" by Millet and son (*JSNHF* 1906)	193
Appendix 5 27 Tree Violets shown at the SNHF by Millet and son, (*JSNHF* 1913)	196
Appendix 6 Bibliography of writings by Armand Millet, including brief notes	197
Bibliography	201
Acknowledgements and thanks	203
Index	205

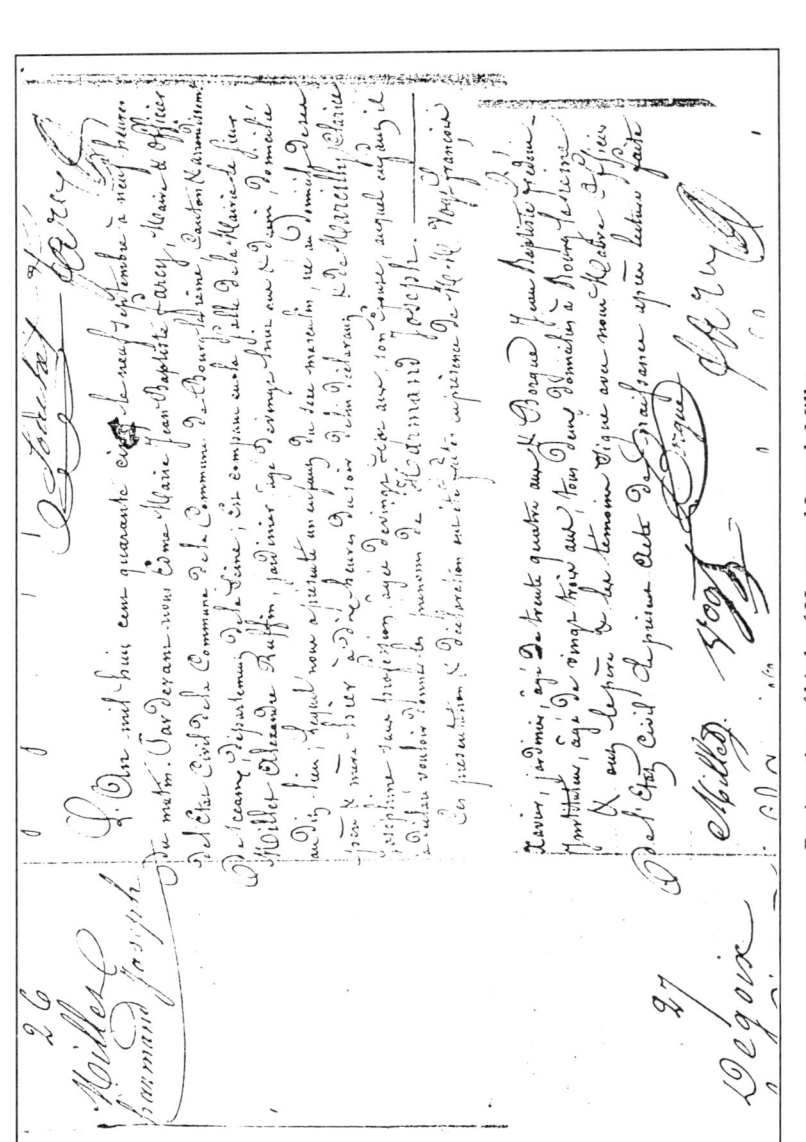

Registration of birth of Harmand Joseph Millet.

ILLUSTRATIONS

	page
1. Registration of birth of Harmand Joseph Millet (cote 4E/BRG_3; courtesy of the Departmental archives of Hauts-de-Seine)	vi
2. Bourg-la-Reine in the 18th c. (from *Bourg-la-Reine* by Lieutier; courtesy of the Bibliothèque-Discothèque, Bourg-la-Reine)	xi
3. The mayor of Bourg-la-Reine reverts to the use of its pre-Revolutionary name. (cote 4E/BRG_1; courtesy of the Departmental archives of Hauts-de-Seine)	2
4. Armand Millet´s family and forebears	4
5. The forebears of Millet's wife, Marie Rosalie Varengue	5
6. A map of Bourg-la-Reine c.1859 (courtesy of the Bibliothèque historique de la ville de Paris)	6
7. Armand Millet's grave	21
8. The church and mairie of Bourg-la-Reine in 1846 and in 1914 (from *Bourg-la-Reine* by Lieutier; courtesy of the Bibliothèque-Discothèque, Bourg-la-Reine)	22
9. The church and mairie of Bourg-la-Reine in 1994	23
10. The title page of *Les Violettes*	30
11. Wood engraving of a violet: from Dodoens' herbal *Florum et coronariarum...*	37
12. The violet *'en arbre'*, thought by Millet to be the same as the 'violette en pyramide'	43
13. 'Princesse de Galles'	68
14. *Viola odorata*: Millet's illustration showing its habit of growth	79
15. *Viola cucullata alba*	95
16. 'Armandine Millet'	124
17. Violet kiosk at Toulouse at the beginning of this century (from *La Violette de Toulouse* by Carré; courtesy of the Royal Horticultural Society Lindley Library)	158

		page
18.	The Marché des Innocents in 1855 (courtesy of the Bibliothèque historique de la ville de Paris)	162
19.	A Tree Violet as trained in the mid-19th c. (*Illustration Horticole*, 1863; courtesy of the Royal Horticultural Society Lindley Library)	185
20.	A Tree Violet trained in the table, or parasol form (*Le Jardin*, 1906; courtesy of the Royal Horticultural Society Lindley Library)	187

PREFACE

In 1961, when I found an English nursery that specialized in violets complaining of their inability to obtain stocks from Millet, I tried to find out what had become of this famous French firm. I knew that about 1930 Lionnel Millet left Bourg-la-Reine and moved to Amilly. I wrote to the mairie at Bourg-la-Reine for information about the original nursery. I received a reply from an official in charge of parks and gardens in the department of Seine, who was based at Sceaux. He informed me that the area where the Millet nursery had been was now built over and he had no further information to give me.

I wrote twice to the mairie at Amilly. On each occasion the mayor replied politely, saying that Lionnel Millet had left Amilly more than ten years before, and that there was now no trace of his nursery or plants. Next I copied down the addresses of nurserymen named Millet and wrote to several, hoping that one of them might be a relation of Lionnel. Clovis Millet of Villemoutiers (Loiret) answered very pleasantly although he had no connection with his namesake. He told me that Lionnel's business had failed. He also gave me the address of Joseph Maréchal at Olivet, as one of the nurserymen with a fine collection of violets. Unfortunately by this time my resolve was weakening and I did not write.

Meanwhile, buying violets from large firms in France, I discovered that their plants were supplied by specialist growers who could not themselves send the violets to me because of the curious – to say the least – way in which phytosanitary regulations were enforced. That is how I came to know Henri Girault at Orléans, and also Marcel Costentin at Saint-Cyr-la-Campagne. Their lists confirmed the idea that all commercial naming of violets was still derived from Millet, including the mistaken 'Rose double de Bruneau' listed as well as 'Rose double' (both Henri Girault and Marcel Costentin informed me that they are identical; in Millet's catalogue they were given as synonyms). One of my pleasantest memories is of being allowed to copy a Millet catalogue of violets in a spacious kitchen while Marcel Costentin went to dig up plants for me and Madame Costentin continued with her preparations for their dinner.

My enquiries at this time, feeble because of limited time, money and resources, had led to nothing, although it gave me the pleasure of a slight acquaintance with those admirable cultivators who seemed to be almost the only people who really valued Millet and his passion for violets.

Two years ago, with more time and money if less energy, I decided to make one more effort. I was annoyed by the way in which Millet's book was referred to as if it were the final authority, but his mistakes were still being quoted by those who should know better. So I set about translating *Les Violettes* and annotating it to the best of my ability.

I have also tried to give a brief account of his life. I hope this will encourage someone to trace all the missing information about Armand Millet: his childhood, his experiences abroad in 1862, his activities in 1870, his personality. Photographs of Millet, his family, his prize-winning exhibits and his nursery; records and documents of the firm – these do not seem to be available. With many nurseries (there is one not so far from Bourg-la-Reine) all this exists in abundance. It is to be hoped that some at least of this material will come to light.

"A tout seigneur tout honneur" was one of Millet's favourite sayings. He himself has not yet been given his due.

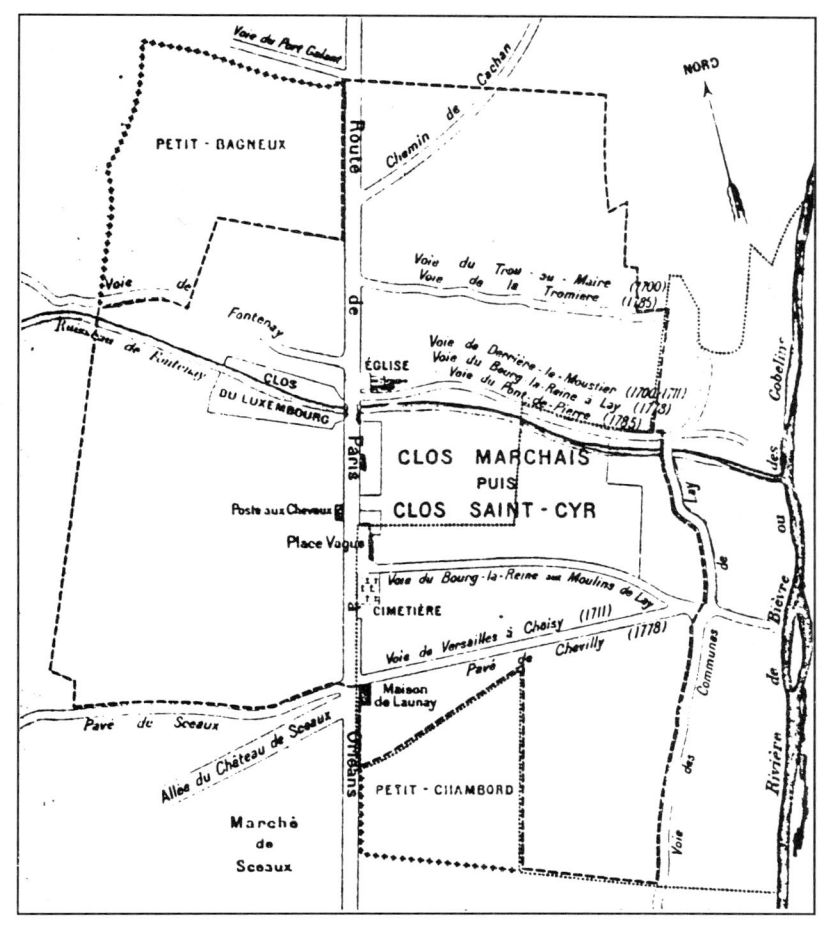

Bourg-la-Reine in the 18th c.

ARMAND MILLET

On the 7th October 1812 Napoleon was in Moscow, but he found time to sign a decree stating that the commune of Bourg-Egalité, in the arrondissement of Sceaux, in the Seine department, should resume its pre-Revolution name of Bourg-la-Reine.

It would seem an unnecessary preoccupation on his part at this crisis in his life, as the matter had been settled more than five years before when the mayor, Lavisé, having used the name Bourg-Egalité (or Bourg de l'égalité) throughout 1806, substituted Bourg-la-Reine on 1st January 1807. But no doubt Napoleon was not satisfied until even such a small matter had been officially pronounced upon.

At that time grapes were grown on a large scale to the south of Paris to supply the city with its vast consumption of wine. There were even many growers of vines in the area whose name was Millet. One of them, Jean-Baptiste Millet of Chilly-Mazarin, had a son born on 28th April 1787 who was called Jacques Saturnin. Throughout his seventy-three years Jacques Saturnin remained at Chilly, and was still working on the land at the time of his death.

In 1810 he married Marie Germaine Baloche, who came from Rungis, and they had several children. One of their sons, born on 28th March 1817, was named Alexandre Ruffin. The rather unusual name Ruffin was taken from a witness at the registration of his birth, Pierre Ruffin Baloche. In his book Armand Millet states that by the beginning of the nineteenth century violets were grown at Rungis on a large scale for cut flowers. His grandmother's family lived at Rungis, so as a boy he must have met people from there, and probably visited it frequently himself.

In 1817, the year of Alexandre Ruffin's birth, the population of Bourg-la-Reine had fallen to 629 and its church was in ruins. When Millet père, as he was to be known, came there about 1840 (Armand tells us his father was growing violets before 1838) there can have been few signs of the

1

thriving centre of horticulture – with such renowned names as Jamin, Durand, Margottin, Delabergerie, Jost, Bruneau, Nomblot – and Millet – among them that Bourg-la-Reine was to become in the second half of the century.

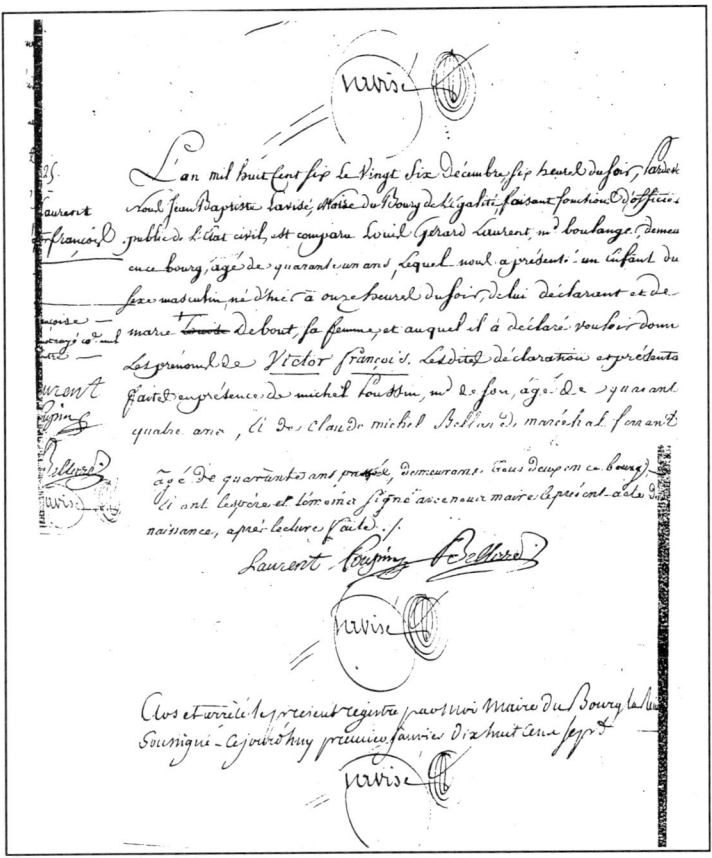

The mayor of Bourg-la-Reine reverts to the use of its pre-Revolutionary name.

Another newcomer to Bourg-la-Reine at this time was François Nicolas Désiré Bréviande, a dealer in brooms, who in about 1828 came with his family from Olivet, near Orléans. It was his niece, who was also born in Olivet, that Alexandre Ruffin Millet married. François Bréviande died in 1839, but his family remained at Bourg-la-Reine, where his widow carried on his business.

On 7th December 1844 Alexandre Ruffin Millet and Clarisse (sometimes spelt Clarice) Joséphine Marcilly, described at the time as a cook living in the fourth arrondissement of Paris, were married at Bourg-la-Reine. Her father, a widower and a dealer in brooms like his brother-in-law, still lived at Olivet. Alexandre Ruffin appears as Carabinier 1st Class and trainee (or employee) gardener (garçon jardinier). One of the witnesses was François Xavier Vogt, an enthusiastic grower of violets on quite a large scale.

On 8th September 1845 Alexandre and Clarisse's son was born, and on the following day he was registered as Harmand Joseph. In later life he regarded the initial H as superfluous, and with the exception of some official documents always used the spelling Armand when not simply signing himself Millet. His father was now described as gardener, and his parents were making their living by growing violets and selling them as cut flowers.

In 1845 Bourg-la-Reine was an auspicious place for a horticulturist to be born. It was a small town with nearly 1500 inhabitants, many of them engaged in market-gardening or the cultivation of flowers. Adjoining it were Bagneux, Fontenay-aux-Roses, Sceaux, Antony, Larue, Chevilly and l'Haÿ. The old 13th century church had been demolished about 1835, the cemetery having been removed to its present situation in 1820. In 1845 a new mairie was built, with the school on the same site, alongside a church of classical appearance, in the new road leading to l'Haÿ. The next year a railway was built from Paris to Sceaux, with a station at Bourg-la-Reine not far from the mairie. The famous nurseryman Jean Laurent Jamin had made Bourg-la-Reine the headquarters of his far-flung horticultural empire and settled there with his father. In 1845 he brought his son-in-law Durand into the business.

Two things encouraged growers to migrate to Bourg-la-Reine and its neighbouring small towns. Nurseries at Paris were beginning to be squeezed out by the demand for land to build on, or driven out by smoke pollution from factories; whilst the shorter distance from the Paris markets encouraged growers from further out to move closer to the capital, and the railway.

The two witnesses to Armand's birth were the young schoolmaster Jean Baptiste Frédéric Borgue and, once again,

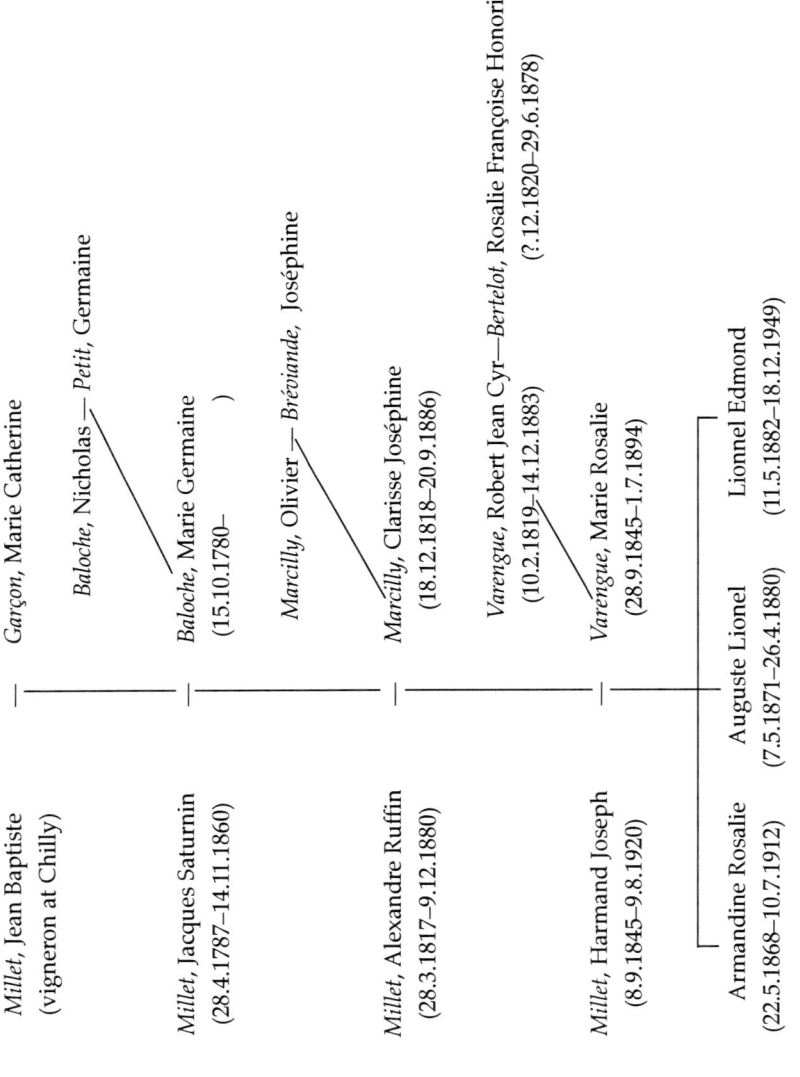

Armand Millet's family and forebears.

Armand Millet 5

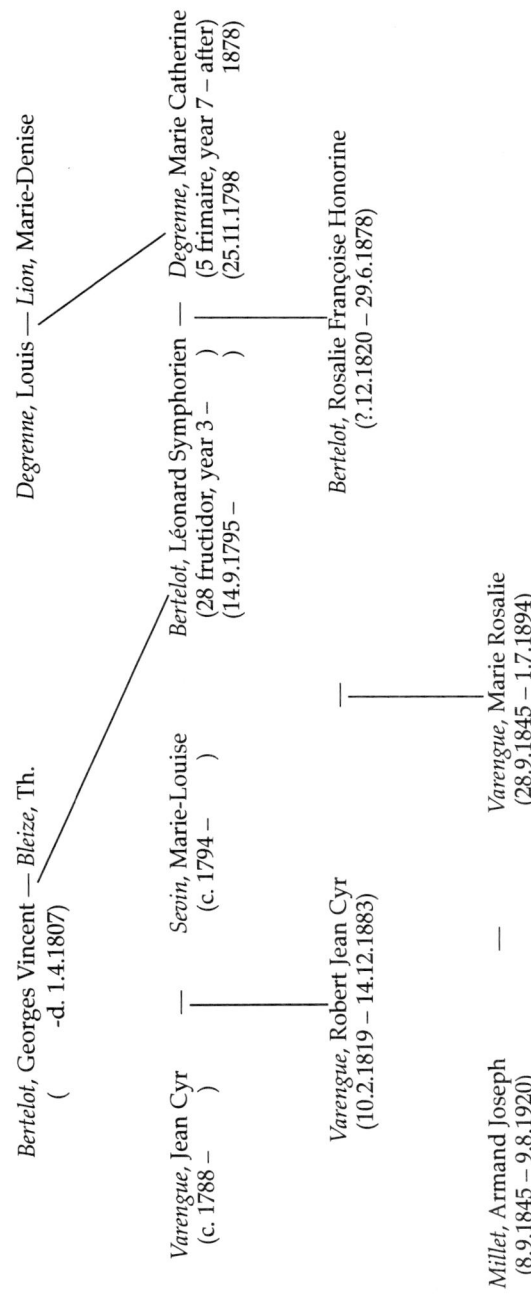

Rosalie Françoise Bertelot was descended from vignerons on both sides of her family. Georges Vincent Bertelot died in 1807, but Louis Degrenne and his wife were alive in 1844. Rosalie Françoise had a younger brother, Léonard Constans. The land she owned at Bourg-la-Reine, and which passed on to her daughter, was presumably inherited from a childless aunt. After her death she was registered as Marie Françoise.

The forebears of Millet's wife, Marie Rosalie Varengue.

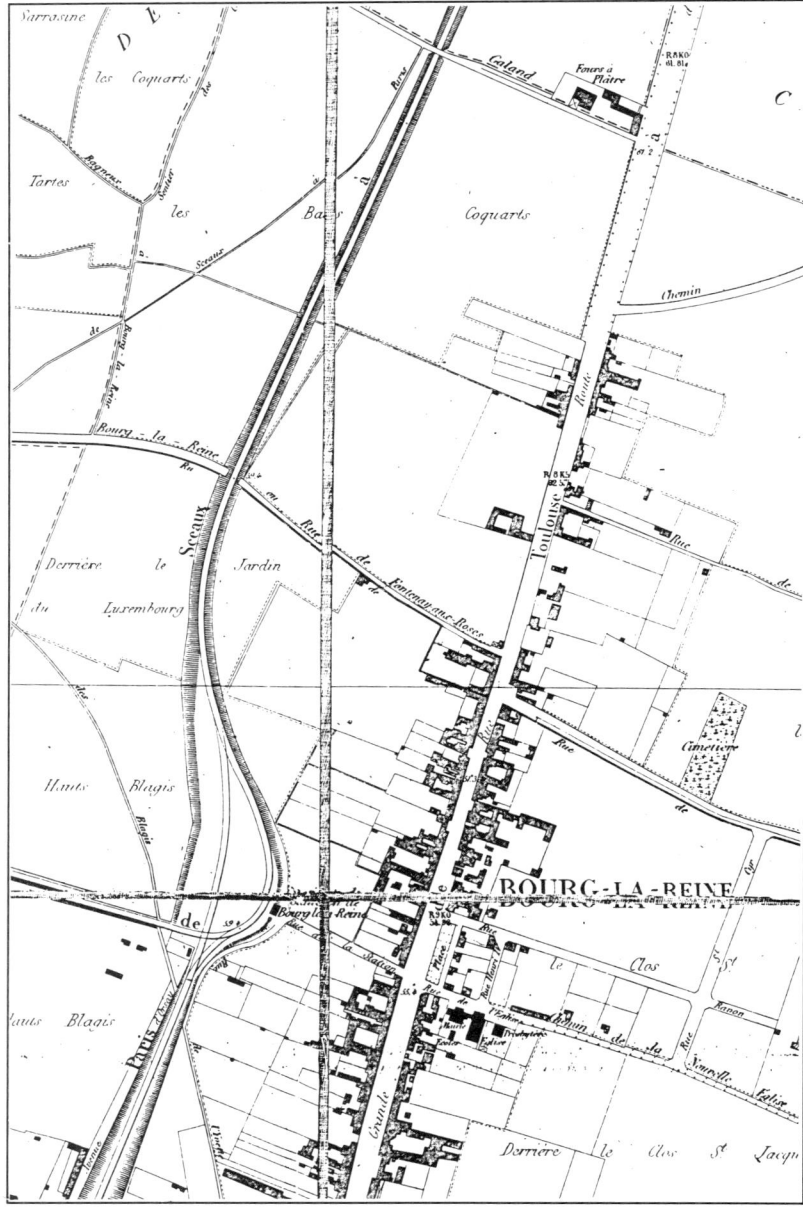

c.1859—The old church has been demolished and the new cemetery is in the road where the church had stood. The railway station is on the other side of the Grande Rue, in line with the new Mairie, school, church and presbytery.

his father's friend Vogt. No doubt Armand made good use of his years at school, perhaps being taught by Borgue, but much more important were the atmosphere of his father's nursery and the skills he acquired there from earliest childhood.

On leaving school, as well as helping his parents, he may have worked at more than one other nursery, gaining as wide a range of knowledge and gardening techniques as possible, in the way that nurserymen such as Bruneau and Lemoine began their careers. In 1862, aged seventeen, he travelled to Belgium, the Netherlands and England, visiting some of the great nurseries of those countries.

The Larue-Chevilly and Villejuif area, adjoining Bourg-la-Reine, contained many nurseries, including the great Croux establishment, which at about this time moved to nearby Châtenay. The brothers Robert Jean Cyr and Antoine Varengue (or Varingue) were at Villejuif, but some time before 1867 Robert Jean moved to Bourg-la-Reine, where he lived with his wife and daughter at 76 Grande Rue. Unfortunately I have not found any information about Armand's connection with nurseries other than his father's, or suggestions as to how he was able to travel abroad in 1862. It is possible that Varengue was in some way involved and that he and his wife Rosalie Françoise Bertelot had been impressed by the young man, since they approved of his marriage to their daughter and duly gave a dowry. On 2nd May 1867 when the marriage took place the couple were both still only twenty-one years old, and Armand had to show he was not liable to conscription.

The following year their daughter Armandine Rosalie was born. In the same year, 1868, Armand succeeded to the direction of his father's business. The father had already had success with violets and strawberries, but perhaps he was feeling the strain of ill health. In any case they seem always to have got on well together, and it gave Armand the opportunity of developing his own nursery.

It was common practice for nurserymen to set up on their own immediately after marriage, no doubt the two acts being part of the same plan. What was unusual in this case was the youthfulness of Armand and Rosalie, and this would seem to indicate that the Varengues were convinced that Armand was a quite exceptional horticulturist. The Millet firm was now headed by father and son, as was to happen later when Armand brought his own son into the business.

Everything now seemed propitious for the development of the nursery when suddenly in the late summer of 1870 disaster struck. The French army surrendered to the Germans at Sedan, and the invading troops advanced to besiege Paris.

Although some people went elsewhere – J.L.Jamin went to Doué in Anjou, where he had part of his nursery – most inhabitants went into Paris as had been arranged. By 16th September almost all had left and settled in the St.Sulpice and St.Germain des Prés district. There the conseil municipal of Bourg-la-Reine met, first at 40 rue du Four, then later at rue de Rennes until 3rd March 1871. The only French people to remain at Bourg-la-Reine were the sisters of St.Vincent de Paul, the orphans they cared for and their caretaker.

Obituaries say that Armand Millet joined the army in 1870. It would seem unlikely that he would leave his two year old daughter in Paris with his wife and the grandparents, and one would assume that the reference is to service in the National Guard. But as his father had served in the army this may have caused him to take some unusual step in September that was remembered later; or it may refer to some incident in which he distinguished himself during the siege.

As is well known all those in Paris were reduced to near starvation towards the end of the siege, which came in late January 1871. On 7th May 1871 a son, Auguste Lionel, was born, not at Bourg-la-Reine but at Longjumeau, close to Chilly, Alexandre Ruffin's birthplace. There had been less destruction in that area.

The boy may have been a victim of the siege himself, as he was less than nine years old when he died.

When the growers returned to their devastated lands and homes in early spring 1871 they faced a daunting task. They had to restore the ruinous state of the land, crops and buildings caused by enemy occupation and undisciplined patrols from Paris. The report of a commission set up to assess the extent of damage suffered by growers, in which Bourg-la-Reine is one of the places named, speaks of loss of plants and tools, looted and damaged houses. With picturesque sarcasm the writer said that the victorious German troops returned home weighed down not so much by their laurels as by French furniture.

In the event only very limited assistance arrived from

elsewhere, but somehow the people managed to get back to normal conditions.

As Millet tells us in his book, he followed his father's advice and grew flowers – violets and roses – rather than vegetables for the following autumn and spring. Despite the severe weather, with the temperature falling to -21° on 9th December, he grew and marketed them successfully.

The last German troops left France in 1873, and Armand must soon have felt that the business was firmly reestablished, for in 1874 he was confident enough to join the French National Horticultural Society.

He became a member on 25th June, nominated by François Lapierre of Montrouge and Athanase Robine of Sceaux, both expert growers of strawberries.

He immediately began to exhibit and take an active part in meetings of the Society. On 13th August he displayed strawberries that had been grown so as to fruit both early and late in the season, and some melons. He gained a 2nd Class award for his strawberries. He also presented a note to explain his methods, and asked the Committee to identify some cherries he had brought along.

That year the Society held a show at their headquarters from 10th to 14th October and Millet entered baskets of strawberries, for which he gained a silver medal and very high praise.

It would be tedious to recount all the awards gained by Millet during his long and successful career, but it is interesting to see how soon he established his extraordinary skill in forcing a whole range of flowers, fruit and vegetables so as to have them in perfect condition for display at the appointed time.

In 1875 he began on 14th January with violets. In March he showed a variety of roses, and started his habit of explaining his methods by stating how he brought his roses into bloom from late November till mid-January and then again from March. In early April he produced cucumbers and roses; in May, at the main exhibition in the Tuileries he obtained a silver medal for strawberries; and on 23rd December he gave a display of French beans.

In 1876 he continued his regular contributions to the meetings of the committees of the Society and again received an award at the main Paris show. He also wrote a note for the journal of the Society – on the most suitable varieties of French beans for forcing.

In spring 1877 he was given the prize of the Vegetable Committee for having gained the most awards – seven. At the principal show, at the end of May, he secured the Society's gold medal of honour for a large display of fruit and early vegetables that included grapes, melon, cucumbers, strawberries, beans etc. And at the autumn exhibition in October he was awarded a large silver medal for a collection of vegetables that included "melons of a rare beauty for the time of year".

Early in 1878, 24th January, Millet showed a number of violets at a meeting of the Society, including 'Souvenir de Millet père', their own variety. The violets and his comments aroused a lot of interest, and he promised to write an article for the journal of the Society. This appeared in April: *A note on various aspects of the cultivation of violets in the Paris area*. It was his first attempt to record the history of this floral industry, in some ways a preliminary sketch for the book that appeared twenty years later.

In 1878 France celebrated its recovery from the effects of the 1870 war by holding an International Exhibition at Paris and, as became the custom on such occasions throughout the world, the main horticultural shows were held in association with it.

The shows were at the Champ de Mars, and at the main one in early June Millet succeeded in gaining the most prestigious award, a large gold medal. He was one of only four in Class 87 to receive this medal (there were five classes in all). Class 87 was for vegetable, or kitchen garden produce, and he entered early vegetables and strawberries.

But there were shows every fortnight from the beginning of May to the end of October, and Millet entered for every one, exhibiting strawberries on each occasion accompanied by a whole range of vegetables. In early August he mounted a display board to demonstrate the usual method of growing melons and explaining his own special practices. This was illustrated by prepared plants at various stages of development and treatment, from flower bud to fully ripe fruit.

At the end of October a tribute to his achievement appeared in the Society's journal: "M.Millet entered the last show as he did the first, that is to say in a way that does honour to our country and especially to our Society, of which he has been one of the most brilliant representatives. His last entry consisted of cauliflowers and strawberries. The cauliflowers were of the finest, in large numbers and of several varieties. They were much admired by visitors, and especially by the English who were showing their produce alongside." The writer went on to say what sacrifices these entries in every show over six months had entailed, as all the fruit and vegetables soon spoiled in the heated atmosphere.

Millet had made his reputation as a formidable exhibitor at the highest level, and this he was to retain for the rest of his active life.

His success and the acclamation he received in early June must have given great pleasure to the family, especially to Rosalie Françoise Bertelot, his wife's mother, who had done much to make it possible, if she was not already too ill to appreciate it. She died on 29th June. So often in Armand Millet's life the satisfaction of success was overlaid by grief, by the death of a member of their close family circle. By her will she gave generous endowments to her daughter that must have secured the expansion of their nursery.

1879 followed the same pattern of success. In March, when showing violets, he resubmitted two varieties raised at their nursery, 'Brune de Bourg-la-Reine' and 'Souvenir de Millet père', and this time they were given a 1st Class award. In the main summer show he gained a gold medal for his strawberries.

The winter of 1879-1880 in the Paris area was appalling, and caused much destruction of plants and trees. On 10th December -23.9° was recorded at Paris. Yet Millet's plans for 1880 were not frustrated. On 12th February he presented a collection of violets that included two new varieties: 'Armandine Millet' and 'Sans Pareille'. The President of the Floricultural Committee said he thought it the largest collection of single violets they had seen and hoped it would be followed by a similar display of double violets, as many varieties of the doubles had become rare if not extinct. In fact his warm appreciation was only marred for Millet by his observing that 'Armandine Millet' had a small flower.

At the June exhibition Millet was again successful with a mixed exhibit of forced fruit and vegetables, being awarded the Minister of Agriculture's gold medal of honour. His name was associated with that of Margottin fils, son of the rose-grower at Bourg-la-Reine, for the excellence of their grapes. In the autumn he was trying to interest people in a potato cultivar to be grown from seed.

It had been a triumphant year in which he had defied the worst known winter frost, proving that frame culture as practised by these skilled growers could produce flowers, fruit and vegetables at will at almost any time of the year. Sadly, although there is no hint of it in the notes and observations attached to his exhibits, it had been a tragic year as well.

His son, Auguste Lionel, died in April; his father on 9th December. There now seemed to be no one who could continue the work of his nursery after him, or anyone to advise and encourage him, as his father, Millet père, had always done.

Millet was now taking an active part in the affairs of the Society: occasionally on a committee or as a member of a commission visiting a nursery or garden; being a judge at or writing a report on a provincial horticultural show; reviewing a book or article for the Society's journal; as well as continuing to bring forward matters for consideration by the Society's committees.

He travelled to flower shows in many parts of France and wrote reports on those at St.Maur-les-Fossées (1881), Caen (1882), Chartres (1885), Nantes (1888) and Périgueux (1889).

In 1881 he undertook to report on a manuscript dealing with potato cultivation. The same year he showed a new variety of potato that was distinguished by its small foliage and was therefore suitable for forcing. In 1887 he was explaining his methods of growing early cucumbers. Until then these cucumbers had been imported from England.

In 1882 what must have seemed almost a miracle occurred. On 11th May a son was born – Lionnel Edmond – who was to become his close associate and carry on the family business after Armand's death. But the small family group he had gathered around him diminished again when his father-in-law, Robert Jean Varengue, died in November 1883. Family associations were important. They provided not only the social pleasures so much enjoyed by people who worked such incredible, as it seems to us, hours of the day, and sometimes night as well; they also sustained each other financially and with their mutual understanding and critical appreciation of their horticultural skills and achievements.

In 1884 he became a conseiller municipal, a local councillor. This must have given him satisfaction, both for the recognition it brought him at Bourg-la-Reine and the opportunity it offered of promoting his ideas on local affairs. Among things he interested himself in was the encouragement of people growing their own produce.

In May 1886 he not only attended the Horticultural Congress at Paris, but contributed a short paper. He had earlier observed that, although it is usually best to sow seed that is fresh, there may be exceptions. He found that seed of melons that was a few years old grew into plants that gave a better crop of fruit, because they had a greater percentage of female flowers, than did those from fresh seed. So it is not surprising that the topic he chose to speak on was: "The effect of the age of seed on plants grown from it", and that melon seed featured in his dissertation.

Despite his fame in the Society as a specialist grower of violets, in the markets for his bunches of violets and among other cultivators for the new varieties he supplied them with, he was presumably not yet widely known to the gardening public as a nurseryman with a good selection of violets for sale. At any rate the *Revue Horticole* noted in 1886 that Millet had issued a list that included a dozen quatre-saisons and other violets, and also double violets of various colours.

In September this year his mother, Clarisse Joséphine Marcilly, died.

Although Millet attended several of the Horticultural Congresses at Paris he does not seem to have given another paper, but in 1888 when the subject of the extortionate (in the opinion of the growers) rates to be charged for freight by railway companies was debated he took a lively part in discussion, and it was the "proposition Millet" that was adopted.

All this time his prowess at the shows had not diminished, but vegetables gradually dropped out of his entries. Fruit in pots, grapes and cherries as well as strawberries, were still a speciality of his: gladioli and perennial flowers were brought in for summer shows: iris, peonies, phlox, sometimes in great variety.

Millet's father had taken a special interest in violets and strawberries, and Armand brought the cultivation of both these plants to perfection. As he had strawberries ready for exhibition throughout the summer he gained countless medals for them, not only at Paris but also in later years at international exhibitions abroad.

1889 was the centenary of the French Revolution, the year of the Great International Exhibition at Paris and the Eiffel Tower. Millet prepared his own sensational exhibit for the principal horticultural show in May. "A collection of two hundred varieties of strawberries in pots, brought to perfection by 'demi-forçage', i.e. being put in frames heated only by the manure in the paths. There were two specimens of each variety. We noticed the abundance of fruit on each plant. We counted as many as fifty ripe strawberries on a single plant".

This spectacular display was long remembered and probably never surpassed.

The same year he wrote two short articles, *Violets* and *Cultivation of Violets*, for a paper he was to be associated with for the rest of his life, *Le Jardin*.

In 1890-1891 he was interested in raising and fixing strains of cyclamen, especially cultivars with semi-double flowers. They appear to have aroused considerable interest at the time, but we do not hear of them later on.

Among the plants that Millet forced for the flower trade was lily of the valley. He experimented with the temperature required and found that less heat than was generally thought necessary gave satisfactory results. In March 1892 he gained a bronze medal for his own exhibit of these flowers.

All this time he had been busy propagating and sending out to growers throughout France violet plants, especially the new large-flowered varieties such as 'Luxonne' and his own 'Gloire de Bourg-la-Reine'. It does not appear that he carried out deliberate crosses then, but he had an unerring eye for the qualities and individuality of a violet, and could propagate these cultivars in very large numbers.

In 1891 he had selected three of his seedlings that he considered to be of outstanding merit. He brought them to the autumn show, but wanted to examine them more thoroughly before naming and launching them. At last he decided to bring them before the Floricultural Committee on 23rd February 1893, when he would name them. As it happened a florist from Hyères, Louis Achard, got in before him by showing at an earlier meeting a violet that appeared to be identical with Millet's No.1 seedling, and named it 'Princesse de Galles'. As with other violets named in France after members of the British royal family, this was because Achard came from the Midi, where Queen Victoria's annual visits made it profitable to pay tribute to her. It is most unlikely to have seemed a suitable choice of name to Millet, even if he had not already decided to dedicate it to his own daughter.

In August 1893 there was a very happy event, the marriage of Armandine to Joseph Henri Gaudichau. The following May their first daughter, Jeanne Henriette, was born. But once again sadness followed closely on happiness when on 1st July 1894, Armand's wife, Marie Rosalie, died. It makes sad reading to be reminded that at the main Show of the Society that year, at the end of May, he had gained six awards, including two gold medals, for a wide range of exhibits.

Octave Doin planned the publication of a gardening encyclopaedia, to appear in small separate volumes and at a moderate price. He asked Millet to write the books on strawberries and violets. For the book on strawberries it was arranged that the general editor of the series, Dr.Heim, should deal with the botanical history and plant diseases. No such provision was made for the volume on violets.

Millet began writing his book on violets in 1895 assisted, as he says, by his own experience and information he had gathered from other growers and above all from what he had learned from his father. But he came across unforeseen difficulties and may have put aside his manuscript in an incomplete state in 1896. As published in 1898 it bears all the signs of a hurried printing that had not been properly proof-read. When received by the Society early in April it was immediately referred for consideration to the committee awarding the Joubert de l'Hiberderie Prize. In the state in which it was unaccountably printed it can have stood little chance. Unfortunately neither Millet himself, nor his son Lionnel, found time to revise and supplement the material it contains. As it is, it is the most important book there is on violets, the only account of the violet industry in the Paris area during the nineteenth century, but it is by no means a true credit to Millet's great abilities.

In 1898 he gained his first public award, Chevalier du mérite agricole, in recognition of his successful career.

His position as an expert on strawberries was recognised when he was asked to contribute an article for the *Plebiscite on Strawberries* appearing in the *Journal de la Société Pomologique de France*. This was in preparation for a congress held at Dijon in September 1898.

In 1897 Millet had become second vice-president of the Floricultural Committee. When, two years later, the president of the committee, J.B.Savoye, died, Millet delivered the funeral speech. It was an impeccable performance. He combined an appropriately rhetorical tribute to the deceased with a simple and sincere account of Savoye's personality and his qualities as a very knowledgeable gardener.

Millet remained a vice-president of the committee the following year but did not continue in 1901. Perhaps he felt he could not spare the time.

We know nothing of Armand's experiences in England in 1862, but they probably determined his plans for his son. Late in 1897 Lionnel went to a nursery at Hampton on Thames, where he spent two years, familiarising himself with English practices in the management of greenhouses and marketing of flowers, vegetables and fruit.

In 1900 Lionel was eighteen and immediately became a member of the Society. Once more the business was Millet and son, with Lionnel taking an active part in affairs. It is not possible to give credit to Lionnel individually for his part in the work of the nursery during the next twenty years, since nearly all exhibits and articles were presented jointly. Their overriding preoccupation with Tree Violets is dealt with elsewhere in this book.

Armandine was living at Neuilly with her husband and two daughters, Jeanne Henriette and Marguerite (born 1897). When her husband died in 1902 she returned with the girls to live at Bourg-la-Reine.

French growers were not easily deterred from showing their wares in distant countries. It comes as a bit of a shock to read of Vilmorin gaining a gold medal at Sydney in 1880, but their display would have consisted mainly of seeds. However, exhibits of fruit and shrubs were sent to shows throughout Europe, including St.Petersburg.

By 1900 French fruit growers were examining the

possibilities of exhibiting in the United States, where refrigerated wagons were being introduced on the railways.

In 1901 a group of Parisian fruit growers, including the famous growers of grapes at Thoméry, decided to send exhibits to the International Exhibition at Buffalo. Leaving Paris on 12th September their wares set off from Le Havre on the 14th on the liner Aquitaine and reached New York on the 21st, ending up at Buffalo on the 24th. Apart from melons and some thin-skinned grapes the fruit was in good condition and was warmly praised. Four awards were given to French exhibits, which were not eligible for competitive classes.

This experience was carefully analysed, and a number of French growers decided to enter the horticultural show at the Saint Louis International Exhibition of 1904. Millet was one of those who on this occasion gained a gold medal.

Millet now exhibited world-wide, sometimes acting as a judge at such horticultural shows. Among those he attended were Milan (1906), Saragossa (1908), Turin (1909), Brussels (1910), Roubaix (1911) and Ghent (1913).

In 1902 a horticultural exhibition was held at Bourg-la-Reine from 6th to 14th September. It was organised principally by Monprofit, a prominent radical journalist, and Alfred Nomblot. It was considered a success, and in 1905 another show was planned, also to be held in September.

This was a grander affair and aroused much interest, as the Minister of Agriculture, M. Ruau, was to visit it. There were complications for the organisers, one of whom was Millet, as the date had to be postponed to fit in with the Minister's arrangements, and there was much relief when he eventually appeared. The two parts of the show, one on the open space between the station and the Grande Rue, the other on the far side of this road, were linked by an elaborate floral arch.

There were many entries to the various classes, and an unexpectedly large number of visitors. On the Thursday, the day the Minister came, two hundred people sat down to a dinner that was prepared by a Parisian firm of caterers. This was followed by the customary speeches, one being by Millet. Afterwards public awards were announced: among these was Millet's promotion to Officier du mérite agricole.

Towards the end of the nineteenth century Millet had begun to raise dahlias. In a short article he wrote for the *Revue*

Horticole in 1910 he tells us that he crossed a dahlia he had bought from a gardener at Montfermeil with dwarf dahlias from his collection of varieties, some of which showed a tendency to having ray petals with a border of a contrasting colour, and also with a sulphur-yellow semi-double dahlia.

The first variety that satisfied him flowered in 1901 and was named 'Paris'. Others that he raised with ray petals having a border of a contrasting colour he named after buildings in or features of Paris, and the whole class he called Parisian Dahlias. Among them were: 'La Seine', 'Notre Dame', 'Obélisque', 'Observatoire', 'Opéra', 'Tour Eiffel' (which was awarded a certificate of merit), 'Tour St.Jacques' and 'Trocadéro'.

In 1904 he began to sell nine varieties of these dahlias, and the following year exhibited this flower for the first time. In 1906 he showed them non-competitively, in 1907 and 1908 he was given a 1st Class award on each occasion he brought along examples of his new varieties. In 1909 he gained a large silver-gilt medal for his dahlias at the beginning of September, and a gold medal at the International Exhibition at Nancy at the end of the month.

He went on showing his Parisian Dahlias with great success until the end of 1913, often displaying dahlias at the same time as his Tree Violets.

The first horticultural show scheduled for the Franco-British Exhibition of 1908, held in London at Shepherd's Bush, was cancelled because it clashed with the date of the arrival of the French President.

A sizable contingent from the French National Horticultural Society came over on 22nd June for the main show. Visits were arranged for them to nurseries such as those of Veitch and Rochford, as well as to Kew Gardens, Hampton Court etc.; and there was a cruise along the Thames. Veitch seems to have acted on their behalf, and at the R.H.S Council meeting on 9th June he promised to give the numbers who would be coming to the fortnightly show on 23rd June, when the French visitors were to be offered "champagne and a biscuit". The number turned out to be one hundred, among them the President and Secretary of the French Society. The minutes for 23rd June record that they were "entertained at a light luncheon in the lecture room".

Millet provided a collection of peonies as part of the

permanent decorations of the Exhibition. In the competitive show on 24th-26th June he gained a gold medal for a display of forty varieties of strawberries (though English reports complained that there were traces of sulphur on some of the fruit). According to the *Gardener's Chronicle* his contribution to the show on 30th September/2nd October was "14 boxes of Strawberries, consisting of perpetual fruiting varieties, some known in this country and others not known. Of these last we may mention 'Mme Bottero', 'La Productive', 'Piex', 'La Perle', 'Oregon', 'Four Seasons (Millet)', 'Merveille de France', 'Odette' etc. Numbers of violets in bunches were also shown".

What would some of us give to see those bunches of violets!

The British gardening press was grudging towards the French exhibitors. What they admitted was a superb display of fruit by Croux they attributed to the superior climate in France; they said the French were given an unfair allocation of space for their exhibits; that these were not skilfully displayed and so on. For their part French journalists said their exhibitors had not taken the show seriously enough and were surprised by the large crowds and the keen interest they took in all that was to be seen.

Millet promptly named one of his new strawberry varieties 'Londres 1908', and it seems almost certain that he must have been among those who came over in June. It would be interesting to know if he did, and if so whether Veitch's nurseries were familiar to him from 1862.

He was deeply involved in the organisation of the French horticultural section of the Brussels International Exhibition of 1910. There was a permanent French exhibit with gardens to which he contributed flower-beds to be maintained, as well as supervising participation in temporary exhibitions. To add to the work a fire caused damage which made it necessary to rearrange some of the shows in the second half of the year. His must have been a stressful task, but he was rewarded by being made Chevalier de la Légion d'honneur.

This was a great honour for him, but it was followed by personal tragedy. On 10th July 1912 his beloved daughter Armandine died; the little girl who had survived the siege of Paris and gave her name to his violet 'Armandine Millet'; who became the mother of his two granddaughters and was commemorated in the strawberry 'Mme.Gaudichau'; and was finally to be remembered in the beautiful violet 'Souvenir de Ma Fille'.

Meanwhile in the summer of this year 1912 an International Horticultural Exhibition was held in the Chelsea Hospital grounds, so giving rise to the venue of the Chelsea Flower Show we are familiar with today. Millet and his son did not fail to put in an exhibit, entering the class for herbaceous plants, perhaps showing iris. It was non-competitive, as no awards were given to foreign exhibitors.

Millet retained his interest in strawberries, and each year he and his son wrote a report on new varieties. This continued until 1914. As had been the case in 1870 there was no thought of war, everything proceeded as normal. But after August 1914 there was little concern for horticulture, except for the production of food, for years to come.

During these dark days, with Lionnel away and work at the nursery reduced to a minimum, nearly all thoughts were concentrated on the war effort. In 1916 Armand wrote: "My son has again been wounded. As he has been in hospitals and convalescing it has not been possible for him to help me in my study of strawberries during 1915."

Nevertheless Millet continued to raise new cultivars of violets. In 1915 he produced a remarkable seedling, the result of a cross between 'Lilas' and 'Perle Rose'. It was not until 1920, when the activities of the Society were at last nearly back to normal, that he and Lionnel presented it, under the name of 'Coeur d'Alsace', commemorating the reunification with France of her lost territories.

From the end of the war Millet, like other nurserymen, had been concerned in helping growers in the ravaged areas that had been occupied or fought over. His thoughts must have gone back to his own struggles in 1871. But he was now a sick man; he had been too ill to stand for reelection to the town council. He died on 9th August 1920.

He had long been known familiarly as "le père la Violette", and Lionnel echoed this association between Armand Millet and his favourite flower when, soon after his father's death, he wrote of: "My poor father, who was born, lived and died among Violets, the flowers he loved so well."

The setbacks he encountered did nothing to defeat his

Armand Millet's grave.

determination. He was fortunate in being provided at an early age with the resources to develop his nursery, but he was brilliantly successful in his use of this opportunity. He

The mairie and church of Bourg-la-Reine in 1846.

The mairie and church of Bourg-la-Reine in 1914.

could justifiably look back with pride at his long active life, at his "modest social position" and the Legion of honour, the varieties of flowers and fruit he had raised, his independent outlook and his concern for others.

His multifarious activities would seem to have taken up all available time, and it is surprising to find him referring to leisure pursuits, botanising in the Alps and shooting game birds in the Pyrenees. Even then he was making practical observations. By any standards he was a man of heroic stature.

Lionnel continued running the nursery at Bourg-la-Reine until 1930. He then moved to the Domaine de Viroy at Amilly (Loiret). There he went on specialising in violets and iris, supplying nurseries throughout France, and many overseas, with their stocks of violets until another war in 1940 seems to have destroyed the nursery and its precious collection of plants. Lionnel Millet died at Brioude in 1949.

Strenuous efforts have been made to collect cultivars of violets, but it seems unlikely that all those that were available from Lionnel Millet in 1939 will be recovered. It is to be hoped that some of Armand Millet's papers and letters will one day be found and published, together with personal reminiscences that may have been handed down to the present time.

There is little in today's Bourg-la-Reine to remind the visitor of the nurseries and the great gardeners who dominated it from 1840 to 1914: Jean Laurent Jamin, his son Ferdinand and his son-in-law Didier Durand; Jacques Julien Margottin (whose only inheritance when he was orphaned at the age of fourteen was his father's razor) and his son Jules; Bruneau, Jost, Rivière, Delabergerie, Nomblot – and of course the Millet dynasty.

Today only the names of some streets remain to recall

The church and mairie of Bourg-la-Reine in 1994.

these shadowy figures: rue Alfred Nomblot, rue Armand Millet, rue Delabergerie, rue Ferdinand Jamin, rue Jacques Margottin, rue Varengue. There is also a rue des Rosiers. But do not be deceived by rue Violette; this is no floral tribute, but perpetuates the name of the worthy Mme.veuve Violette, who in 1853 bequeathed money to build the sturdy presbytery that stands beside the church.

THE NURSERY

Unfortunately I have not been able to find a description of Millet's nursery. It was on land on the northern border of Bourg-la-Reine, land that had only become part of the commune in 1834. This northern boundary adjoined the Voie du Port Galand (now the rue du Port Galand) and Millet rented from the Bourg-la-Reine council a strip of land along the road verge as far as the railway line. An access road for the nursery and greenhouses can be seen on a map of the 1890's.

This area acquired by Bourg-la-Reine had been known as Petit-Bagneux. The railway line from Paris now ran through it, forming the western boundary of the nursery. Its eastern limit was the Grande Rue, and the southern boundary was probably where at present rue Armand Millet stands.

Under Armand's direction the nursery contained greenhouses for vines and forced fruit, and for roses to provide flowers in winter and early spring. They also gave him the opportunity to experiment with new techniques in growing out of season fruit and vegetables in pots.

A large part of the nursery would have been taken up by the 'jardins' with their frames where the bulk of the violets and strawberries were grown, as well as other flowers and vegetables, if they needed protection or heating from the hotbeds formed around the frames.

There would have been an area corresponding to a 'jardin fleuriste', where most of the propagation and growing on of plant material, raising seedlings etc. could be carried out under constant supervision. And there was an extensive area devoted to those perennials – iris, peonies, phlox as well as gladioli – that he exhibited in such a large number of varieties. There is a photo of this part of the nursery, with their house in the distance, on a postcard possessed by Philippe Chaplain, director of archive material at Bourg-la-Reine.

EXHIBITING VIOLETS

Armand Millet and his father were already known as outstanding growers of violets before Armand joined the Horticultural Society in 1874. He at once began showing plants. The next year he continued with an exhibit of violets in January, to be followed by violets and roses in March. However, these flowers did not play a prominent part in his plan to gain recognition for his skills, and awards to add to the prestige of his nursery. It was not until January 1878 that he brought violets before the Floricultural Committee again. On this occasion the interest the flowers aroused led him to write an article about the cultivation of violets in the Paris area, and this was published in the Society's Journal the same year.

After this his displays of violets became an annual feature of the Society's meetings, with the possible exception of 1881 and 1886.

He also exhibited violets from time to time at shows held in spring or autumn. For example he showed 'Luxonne' in 1888; it was at a show in 1891 that he displayed for the first time his three seedlings, one of which was to become the most famous of all single violets worldwide, 'Princesse de Galles'; and in 1896 he showed 'La France'.

He gained many medals for violets at these shows, including bronze in 1887, silver-gilt in 1892, and his first gold medal for this flower in 1896. He also judged violets at shows from time to time (e.g. 1890, 1897) and although this meant his own violets could not compete for an award this did not stop him from bringing along colourful exhibits to add to the occasion.

He was almost the only person to display violets at meetings of the Floricultural Committee, so when on 9th February 1893 the florist Louis Achard preempted his claim to what Millet regarded as his own new violet and named it 'Princesse de Galles', Millet must have had a most unexpected and unpleasant shock on hearing about it.

Perhaps it was as well that Armand was not present himself on that occasion!

What made these displays so entertaining was their variety. Although he often gained awards and commendation for his violets and his cultivation of them, this was not his sole purpose. And as it was Armand who suggested to Claude Néant that he should take along his seedling Parma violet (later to be known as 'Mme Millet') it was he who was responsible for the discomfiture of the Committee when Néant brought his pot with a Parma violet bearing a seed capsule, thus dramatically disproving their assertion that this was impossible.

One of his principal concerns was to show a variety of cultivars, both old and new ones, but especially those that came from his own nursery or that he was launching.

He presented a succession of new or recently introduced varieties right up to the end of the century. These included 'Brune de Bourg-la-Reine' and 'Souvenir de Millet père' (1878); 'Armandine Millet', 'Lilas', 'Sans Pareille', 'White Czar', 'Sans Prix' (1880); 'Violette dite de Bruneau',[1] 'Mme.Millet', 'Roi des Violettes' (1884); 'Gloire de Bourg-la-Reine', 'Swanley White' (1887); 'Mme.E.Arène' (1891); 'Amiral Avellan', 'Princesse de Galles', 'Explorateur Dybowski' (1894); 'La France'[2] (1896); 'California', 'Mme.E.Dutertre', 'Princesse de Sumonte' (1897); 'Mlle.Reine Augustine', *V.od. sulphurea*, *V.palmata*, *V.pubescens* (1898); 'Mme.Pagès', 'M.Astorg'[3] (1900).

In 1897 he also showed 'Patrie', which he thought was a very old variety and which he tried to identify with the pseudo-la Quintinye's Tree Violet. It is about this time we first hear of Millet's Tree Violets, and much of the energy of Armand and his son went into the production and development of them.

In 1902 there was a fine display that included the double rose under the name 'Rose double de Brunaut', 'nana compacta' (a Millet variety that did not send out runners), 'Bleue de Fontenay' and six violets that had been brought

---

1. Quite unusually Millet was shown to be mistaken in this plant. Burelle, a member of the Committee, pointed this out and we hear no more of this single violet.
2. In 1897 'La France' was awarded a large silver medal.
3. This quaint version of the name Astor was perpetuated in France.

from Savoy under the names *V.subcarnea, V.collina, V.sepincola, V.od.leucanthera, V.permixta,* and *V.od.purpurea.* Then in 1907 he showed some varieties for autumn flowering. In 1914 came a very important variety of his own raising, 'Souvenir de Ma Fille', commemorating his daughter Armandine, and with it four other violets from their nursery: 'Helvetia', 'Mlle.Garrido', 'Marietta' and 'Rosea delicatissima'. In 1920 Millet and his son showed Armand's last cultivar, the splendid product of his labours during the dark years of the war, 'Coeur d'Alsace'.

At times he brought along collections of different violets: nine varieties (January 1878); silver-gilt medal for a collection of violets (March 1892); twenty pots of old and new varieties of violets (March 1895); fifty varieties of violets that scented the air (April 1900 at the International Exhibition). On other occasions he mounted large displays consisting of just one or two varieties.

In 1877 Burelle had said that the Floricultural Committee was then carrying out research into the origin of cultivated violets, and his remark no doubt prompted Millet into writing his article for the Journal the following year. Whereas we hear nothing of any information gathered by the Committee, Millet collected what knowledge he could, and the violets themselves. Many of those kinds that were not grown in their millions for sale in bunches would have been hopelessly confused had this nursery not supplied them true to name to nurserymen throughout France and many countries of Western Europe, for whom the Millet catalogue was their reference book.

He occasionally found time to bring violets to the Committee in order to demonstrate some point of interest. In March 1884 he showed a number of different Parma violets in pots in order to illustrate his contention that the 'Parme de Toulouse' only differed from the Parma violet usually grown at that time as a result of the place or situation where it grew. In February 1887 he endeavoured to show that 'Mme. Millet' and 'Swanley White' were hardier than 'Parme ordinaire'.The following month he brought along two plants, one of 'Mme.Millet' the other of 'Swanley White', each of which had one flower whose colour was violet. This proved that, though very rare with violets, mutation could occur. And in November 1888 he put before the Committee two plants of

Parma violets bearing seed capsules, perhaps to remind them of their arrogant and dogmatic views on the subject in the past. He said on this occasion that he had offered five francs to any of his employees who should find seed on a Parma violet, but that no one had succeeded in doing so. Only he had spotted these two among the vast numbers of plants being grown.

LES VIOLETTES

LEURS ORIGINES, LEURS CULTURES

PAR

A. MILLET

Horticulteur à Bourg-la-Reine (Seine)
Membre de la Société Centrale d'Horticulture de France
Vice-Président du Comité de Floriculture
Grand Prix et Premier Prix en 1878-1889

AVEC 23 FIGURES DANS LE TEXTE

PARIS

OCTAVE DOIN	LIBRAIRIE AGRICOLE
ÉDITEUR	DE LA MAISON RUSTIQUE
8, PLACE DE L'ODÉON	26, RUE JACOB, 26

1898

VIOLETS

Their Origins and Cultivation

by

A.MILLET

Nurseryman at Bourg-la-Reine (Seine)
Member of the Central Horticultural Society of France
Vice-President of the Floricultural Committee
Winner of Major Awards 1878 – 1889

With 23 illustrations in the text

Paris

Octave Doin | Agricultural Books
Publisher | La Maison Rustique
8, Place de l'Odéon | 26, rue Jacob, 26

1898

TO THE LADIES WHO READ THIS BOOK

It is to you, Ladies, that I have the honour of dedicating these lines. Through your refined taste, from your need for all that is beautiful, your love of flowers, and in particular of those newly raised flowers that increase in loveliness day by day, you have prompted humble gardeners to press ahead, to create and achieve the impossible in order to please you. And so, under your potent spell they have become artists of floral creation; and as it is partly due to your inspiration that I am able to write these pages, permit me to offer them to you as the modest and respectful token of my gratitude.

INTRODUCTION

At first sight singing the praises of something one is very fond of would seem to be a simple matter. But when the subject is of a refined nature, and conceals rather than displays itself, then the work becomes much more arduous. That is what has happened to me in writing this book.

It was my dream to write the life and history of violets, and I had long been requested to do so by numerous friends who were violet enthusiasts, when M.Doin[1], the publisher of works on agriculture, suggested I should turn my dream into reality by producing the volume: Violets.

In doing so he was attacking my weak side. Just think, dear readers: to write the life of the flower one loves is in a way to write the life of those who are dear to you, it is to write the Past; it is also the Present, and even the Future, for those who can combine practice and theory.

And so I set to work on the book in high spirits. But in everything there is always another side of the coin, and I very soon realised that there was nothing – or at least very little – written about this plant. I was, then, left to my own resources, for no lover of violets had yet tackled this subject. I was well aware that there were some accounts of violets and anecdotes in circulation, but all less firmly based on facts the one than the other, without precise detail, and even without any great basis of reality. Called upon by this work to make all possible enquiries into the early history of my protégé, I have been struck during the course of this research by the paucity of existing documents, old or new. Previous to a short article that I published in 1878[2] there was nothing on methods and practice of cultivation.

Though the history of violets was scanty, I was fortunate enough to possess an old work with wood engravings, published in Latin at Antwerp in 1566[3]. I have drawn on it extensively. Our great la Quintinye (1690-1730)[4] also published some ideas on cultivation, and cited several varieties; quite elementary no doubt, but it already marked a step towards the future. This was a second source which I have drawn on a great deal. Then I have been assisted, enlightened and inspired by the best observations, the most precious historic memories, that is to say the experience and the tradition I derive from my forebears; accounts that are at once authentic and practical, to which I have added, in order to present them to my readers, the steadfast exertions of my long career.

I thought I should divide my work into two separate parts. In the one I deal with the historical aspect. In this I cause to pass before the eyes of my readers in succession the old varieties, what they had to offer, and still offer; then those of our own time, how they came into being, why they are always in demand, defying with impunity the proverb that says: Everything wearies and everything passes away (they pass away, but are never wearisome).

In the other part I have applied myself to making widely known the methods of cultivation, in a way that is practical, easy to carry out, inexpensive and within the reach of all. I have done my best to explain the cultivation clearly and give all necessary instructions as simply as possible. To convey a better idea of the appearance of the flower I thought of using photography, which alone can give an exact impression. I carefully photographed the most meritorious varieties. M.Doin, our devoted orchid lover, has spared no pains over the engravings and blocks, which have been done in masterly fashion. Finally, I have tried to prove that with a little effort, of course, you can produce something very fine[5], and that at little cost you can combine the pleasurable and the useful. Such has been my aim during the course of this work. Have I achieved it? That is what I desire most of all. May all my readers find here what they are looking for and what they enjoy, for that will be my finest reward!

              Millet

Bourg-la-Reine. 17th February 1898.[6]

FIRST PART

Historical Account of Violets; the Author's Observations

Dear readers, please excuse me if my subject is not always entertaining, and if at times you notice in this little work a partiality for my protégé. Is this due to my having been born amid great numbers of bunches of violets? A little, probably, but also to the fact that I am indebted to this flower for the modest social position that I occupy. If I allow myself to conjure up in a few lines these memories of the distant past, it is to make it known that from father to son we have earned our living by growing violets, and that every day we were constantly caring for this plant, which was part of our life. God knows, we gave it our best attention, heated and pampered it so as to obtain beautiful flowers, and that at the most difficult times, in the harshest winters. What efforts to achieve improvement of the stock! What trouble taken without success! How many years spent without achieving anything! But, not disheartened, we struggled on (I say we because my father had already obtained several varieties markedly different from the ordinary violets), and I am happy and proud today of the well warranted results we have had. For, despite its modesty, the violet is assuredly the flower which, with the exception of its great sister the rose, is sold in the largest quantity and for the greatest amount of money, especially in France.

Which of our ancestors could have thought that in the year of grace 1895 violets would travel in upholstered and heated railway carriages, like our finest ladies? Not one would have imagined that this little flower, so bashful, growing in our hedgerows, in our great forests, or hidden under the grass of our meadows, would create for itself a prominent position among its rivals of today. What is the reason it has always remained in fashion? That it has the preference above so many other flowers? Would it be because some poets of times past have sung its praises? Because some great men, like Napoleon 1[7], had a fondness for it? I do not think so. My firm belief is that it owes the preference in which it is held to the good taste of French women, who see in it a modesty of attire together with unequalled fragrance, which perfumes our dwellings without, however, being at all disturbing. It makes us think of our transitory life and seems to say: "I am, I pass away, and yet life is good".

Its history stretches back into ancient times. For centuries it grew in the wild, it was gathered on the roadsides, along hedgerows, in meadows. From time immemorial the flowers had been dried and were used, as were the roots, for making alleviating medicines. Nowadays it is only used as a herbal tea (tisane).

There are many legends about its origin and its name. As I have already said, many ancient writers mentioned it, and here are a few traditional beliefs that I have taken from them.

Legends about, and Works on, the Violet[8]

The Greeks named the violet *Ion purpurium*, that is to say purple violet. Theophrastus[9] called it *Ion melan*, that is black violet, because of its colour, purple looking almost black[10].

Pliny[11] says that the word *Ion* indicates the violet, and distinguishes it from species of similar colour[12].

The herbalists have retained the Latin word *viola* and call this plant violet or mother of violets.

The Germans (Germani) call it *Blau veiel* or *Mertzen violen*, the violet that flowers in March.

The French (Galli) – *violette de mars*.

The Belgians (Belgen) – *violetten*.

In his description of the earth Nicander[13], according to Hermolaüs[14], thinks that the Greeks called the violet *ion* because the nymphs of Ionia were the first to offer it as a present to Jupiter. Others asserted that the name *ion* comes from the fact that when Jupiter changed the young Io, whom he loved passionately, into a heifer, the earth produced flowers that sprang up from beneath her feet[15]; and that these flowers took their name *ion* (violet) from that of the girl. For this reason the word for violet in Latin (*viola*) has the same meaning as that for heifer[16].

Similarly Servius[17] says that in Latin the word *viola* (violet) has the same meaning as *vaccinium*[18], and quotes with regard to this the passage from Virgil's *Eclogues* which says:

Alba ligustra cadunt, vaccinia nigra leguntur
(The white privets fall, the black whortleberries are picked)[19]
(The white clematis falls, the dark violets are gathered)
Pallentes violas et summa papavera carpens
(Gathering the pale violets and the tall poppies)

However, Virgil[20], in his *Eclogue X*. shows that the words *viola* and *vaccinium* refer to different flowers:

et nigrae violae sunt et vaccinia nigra
(violets too are dark and dark are hyacinths)

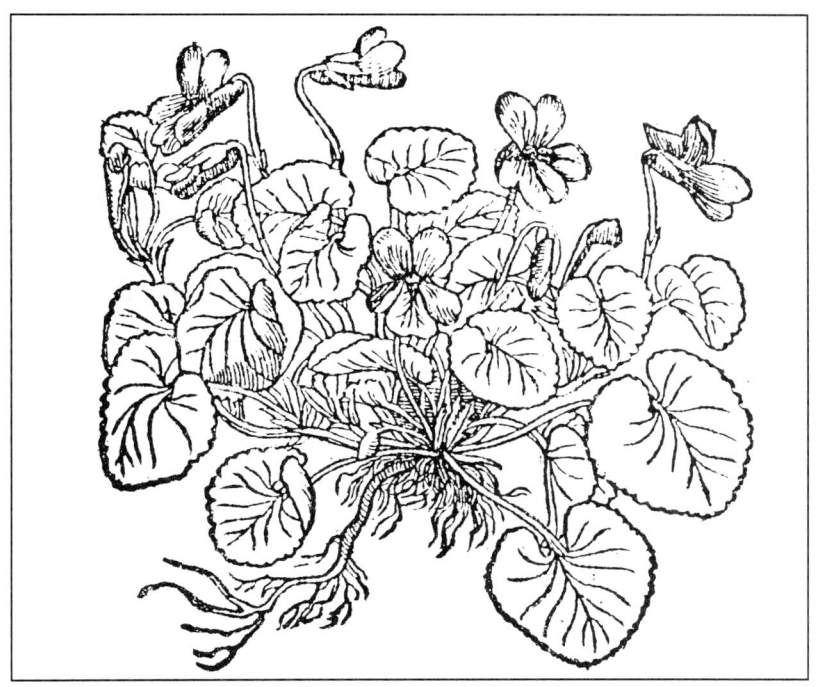

Wood engraving of a violet from Dodoens' herbal *Florum et coronariarum...*

Vitruvius[21], in the seventh book of *De Architectura*, distinguishes the violet from the whortleberry[22], for he informs us that the colour red (red or Attic ochre)[23] is made with violets, and a fine purple colour with whortleberries.

"When the dyers[24]," he says, "wish to imitate Attic ochre, they put dried violets into a receptacle, boil them in water, and while it is still lukewarm pass it through a cloth, squeezing it out with the hands. They put the water that has been coloured by the violets in a mortar, and pouring Eretrian earth[25] into it and pounding it, they obtain the colour of Attic ochre. By treating whortleberries[26] in the same way and mixing them with milk they obtain a fine purple colour."

Kinds of violets spoken of in 1566
(taken from a work in Latin published in Antwerp by the doctor Rembert de Dodone)[27]

Let us leave these far distant times.

I have before me a description of this plant from the year 1564[28], published in Latin at Antwerp by the doctor Rembert de Dodone. In this book, which is full of interest and written in Greek and Latin, the author examines the various fruits[29] and flowers of the time, giving their names in German, Flemish, French and Italian[30]. This book is illustrated with very fine wood engravings, in which I have had the good fortune to recognise many plants after the lapse of 330 years.

It is the first book to mention double violets[31], without however specifying whether they belong to the race of Parma violets. I have good reason to believe that he was referring to ordinary double violets, double blue and double pink[32], which occasionally bear seed. Here is the exact description he gave, without the addition or omission of anything at all.

Of the black or purple violet

The black or purple violet sends out from its root numerous leaves, broad, veined and lightly scalloped, which are thinner, rounder[33], darker than those of ivy, especially on their upper surface or above. From amidst these leaves come slender stems[34], each bearing a single flower, beautiful, sweet-scented, of a deep purplish-blue[35], rarely white, composed of five small petals, the lowest of which is the biggest. The calyx of this flower hangs down and, when it is ripe it forms three compartments; its seed is small and a slightly elongated sphere in shape.

The violet, despite its thin fibrous roots, is herbaceous summer and winter; it has very good resistance to the cold in unexposed positions, beside briar bushes, walls, hedges, in gardens and on the margins of fields; it is abundant, and forms clumps in a rich soil. It usually flowers in March, and at the latest in April.

Belonging to this genre is a double or multi-petalled violet, both purple and white[32], which are only found in gardens.

There is also a kind of wild violet with smaller leaves, paler flowers, with little or no scent; it grows in shady places, along hedges and ditches in land that is usually dry and infertile.

The flowers and leaves of violets have quite efficacious refreshing and emollient properties.

Their flowers are employed for all internal inflammations, especially of the sides and lungs[36]; they alleviate congestions and lesions of the chest, the trachea and the throat, lessen excessive inflammation of the liver, kidneys and bladder, moderate the heat of burning fevers, combat the bitterness of bile, keep away and soothe thirst.

As regards the drosaton, or serapion[37] as Actuarius[38] calls it, which is made by boiling water in which a certain quantity of fresh violets have been steeped, it assuages the abdomen and causes bile to be ejected[39] through the fundament when three or four ounces are taken[40].

An oil which possesses a very efficacious cooling and emollient virtue is made from violets. When rubbed on the temples it induces a very gentle sleep, that is prevented by a hot or malignant fever; mixed with yolk of egg it alleviates the pains of haemorrhoids and the fundament; it is added beneficially to cooling poultices and those that anaesthetise pains.

Whether the oil in which the violets are macerated be made of unripe olives, which the Greeks call omotribo[41], or sweet almonds, as Messues[42] instructs, the violets must be recently picked and fresh; for violets that have lost their moistness are not only less cooling but also appear to have acquired some heating power[43].

It is generally considered that dried violets can usefully be mixed in with medicaments that are regarded as invigorating.

Violet leaves taken in a salad refresh, moisten and soothe the belly.

Applied externally they clear up all inflammations, either by themselves or through the flour of the poultice.

According to the testimony of Galenus[44] they may be applied for burnings of the stomach or inflammation of the eyes.

Dioscorides[45] prescribes them for the breaking out of the fundament[46].

Pliny says that by wearing a garland of violets or by inhaling their scent you can avoid migraine[47] and sluggish feelings in the head. Taken in infusions they prevent attacks of quinsy.

The purple corolla of the violet, taken in an infusion, is a cure, with children, for epilepsy.

Violet seed drives away scorpions[48].

Discorides says that it is not the purple flower of the violet, but *Aster atticus*[49] which, taken in an infusion, effects a cure of quinsy and epilepsy with children.

One may then, in accordance with this author, question the peculiar quality that Pliny attributes to the violet.

It can be seen from this extremely interesting account that Dr.Rembert had been especially concerned with violets from both the botanical and the medical point of view. But he only speaks generally of the violet, without mentioning distinct varieties and without seeking to find out whether this plant possessed different qualities. In any case it is established in his book that violets were not cultivated[50]. Double violets only were grown in gardens. I am convinced that for a long time it remained the same; for from 1566, the date of Dr.Rembert's work, up to 1690 I do not see anything appearing in writings that makes the least mention of the cultivation of violets. I have indeed found a few notes or anecdotes speaking of bunches of violets, but with no description of varieties, and everything leads one to suppose that they came from the woods.

Violets are cultivated in 1690
Some notes on this cultivation by de la Quintinye

In 1690 de la Quintinye, director in chief of the Royal Gardens at Versailles, makes some mention of violets in his instructions on gardening. But in 1730, in his new edition, details are more precise. He shows the way plants are propagated, speaks of double and single violets, red, blue and white; he also cites the tree violet[51]. His description is just the same as it would be given today, although written a hundred and fifty years ago. However, one finds no mention of Parma violets. Since he describes the tree violet for us, if he had known the Parma violet he could not have failed to mention it, as the foliage, the flower and the scent are all different from that of other violets. I stress this point, because I have looked for, and am still looking for the origin of the Parma violet[52] without any success; any more than I have discovered the origin of the three varieties of the hardy sweet-scented doubles – red, white and blue – all three very ancient. There is a gap here which is not at all easy to fill. No flora has been found, as far as I am aware, that mentions double violets growing in the wild[53]; no botanical excursion speaks of them. On the other hand several floras and botanical works say: They become double in gardens as a result of cultivation[54]. Proceeding from this principle (which I can only with difficulty accept) how could it be that they are so ancient; that they go back to a time when violets were not grown in gardens? And then the tree violet spoken of by de la Quintinye, which has flowers so packed with petals, and which then as now never produced seeds – what is its origin? So many unanswered questions! One can make suppositions: that is always easy, but finding proof to support them is quite another matter; and desiring to prove without being sure, that's too bad.

Let us come back to what some writers have said. "Violets become double as a result of cultivation in the garden." This assertion seems to me rather a bold one, and for this reason. As I have already said, four sweet-scented double varieties are very ancient; that is to say, their origin goes back to the most remote times, and when there was no thought of cultivating violets. It must be admitted that it would be a piece of good fortune if double flowers appeared under these conditions. If it is retorted: Why not? refutation is easy and grounded on the following weighty evidence.

I may say that from the beginning of the century from father to son we have grown violets and observed the developments that occurred. We were engaged more particularly in growing violets for cut flowers. For this cultivation there was no choice; there were only two varieties[55], forms of the woodland violet (*Viola odorata*) under the name of quatre-saisons[56] violet. Now, despite very careful cultivation, all of which had one objective in mind, that is to say to get large flowers, not a single grower obtained so much as a hint of doubling in the plants he grew from seed. Yet violets were grown from seed in great quantity. At Bourg-la-

Reine alone between five and six hundred thousand quatre-saisons violets were forced each year. In the surrounding area they were grown in the open by the million, and all these violets were propagated by seed. I consider then that examples of this order are very convincing, and give no certain origin for these double violets.

I have near at hand several novels of times past, but written since the beginning of the century. As the Parma violet was known at this period, it was no more difficult to say Parma violet than Viola odorata or any other variety; and for greater clarity I put before my readers' eyes the few lines of de la Quintinye, without adding or omitting anything whatever, even respecting the spelling of the time.

Violets – The doubles as well as the singles, and of whatever colour they may be, – he wrote – although they produce seed in small reddish vessels, nevertheless they only increase by means of shoots that they send out, each rooting plant growing gradually into a large clump, which divides into several small ones, and these being then replanted become in time big enough to be in their turn separated out into several other small ones.

From these few lines we see that de la Quintinye, while not dealing with propagation of the violets from seed, assures us that they bear seed in little pods. This clearly gives us to understand that he was not acquainted with Parma violets, since those he spoke of produced seeds, and Parma violets never do[57].(1)

(1) I say never, and yet this species does produce seed sometimes. Twice during the period of thirty years I have managed to gather seed, and then the capsules were imperfect; however, the seeds were viable. It happens so rarely that one can say hardly ever. We have never found seed on the 'double blue' or the 'tree violet'.

In another paragraph de la Quintinye says:
Violets, especially double violets, provide pretty borders in our kitchen gardens. Their flowers make a remarkable embellishment when skilfully placed on spring salads. They are increased by dividing large clumps to make small ones.

Finally in Chapter 53[58] he says explicitly:

Of the double violet. – The double violet that is grown in gardens is similar to that which grows wild in the fields, except that the latter has single flowers, while the former has double flowers, white or red or violet, or of several other colours. They all spread, sending out runners. They require a moderate amount of sunshine, and good, strong soil. In hot weather they need watering. They are better in pots than in the open ground, since they can then be put under glass in winter. As they do not bear seed, runners are removed, and these are replanted separately.

As regards the violet in pyramid form[59], it is also called the tree violet. It sends up one or more stems which become covered with a large number of small buds from the base to the top, in the shape of a tall pyramid. These buds, which are elongated and grooved, become larger and are like so many little blue stars, from the middle of which a whitish thread rises. These flowers smell like storax[60]. This plant is worthy of

The violet 'en arbre', thought by Millet to be the same as the 'violette en pyramide'..

consideration, because it is sometimes in bloom for more than six months.

It requires a moderate amount of sun, a good strong soil; it must be watered copiously. It does not set seed, but is increased from the roots, which are full of a milky substance. These roots are broken up into pieces, which take root, send up new growth and produce flowers.

It follows from this last page clearly and explicitly that, as in the time of Dr.Rembert, that is to say in 1566, there were single violets and double violets in three colours: blue-violet, white and red; and also, without any doubt, the tree violet. Now concerning seed, it seems that in the way of double violets de la Quintinye had before him only the double blue and the tree violet. These two varieties are the same as the Parma violet; on no occasion have I ever seen seed on it, and the flowers are so double that no capsule could form on the calyx. This is not the case with the two varieties, double red and double white; they commonly provide themselves with a great many capsules containing seed. This is due to semi-double and single flowers which grow on these two varieties at the end of the flowering season. As with the stock, when this seed is sown more single than double flowers[61] are obtained.

Varieties grown at the beginning of the eighteenth century: development in the South of France

However it may be, it has been established that at the beginning of the eighteenth century there were grown in gardens: 1) the single sweet violet; 2) the double blue, double rose, double white, and also the tree violet, which is a double blue too. This state of affairs was unfortunately to continue for a very long time to come. Since there was no trade in violets as cut flowers, nobody thought of improving the stock. To provide flowers, whatever was at hand was planted in a border. Some people distilled violets in the spring to make domestic perfumery; violets were dried for medicinal use. It was medicine and above all perfumery that got the cultivation of violets going. During the sumptuous reigns of Louis XIII and Louis XIV the display of luxury led to a greater demand for perfumery. About 1750 some families in Provence were already starting to dry violet flowers for pharmaceutical products. They were gathered in woods and by hedges. Then violets were introduced into gardens. Their cultivation was quite primitive. Plants were taken from more or less any place, planted out, and all the runners or stolons were left to develop and fill the beds, thus forming little fields of violets. The years passed. 1770 saw the arrival of several people who purchased land, intent on working seriously at the cultivation of violets.

From then on things got going, fields extended, and from 1780 cultivation increased. High quality perfume was on sale. What had been earning only small sums of money became the source of considerable profit, and from the beginning of the nineteenth century violet cultivation expanded greatly. Products were dispatched to the Orient, in the manufactured form as well as raw material[62], that is to say the flowers themselves.

Violets were grown principally in the district of Saint-Roch, later on at Villefranche-sur-Mer, then in many of the communes of the Grasse arrondissement[63]. They were grown under orange trees; the method of cultivation was quite primitive and of the simplest. As, moreover, is still the case with some old-fashioned growers, they planted out their violets for six, eight and ten years. The violets put out stolons or runners, all of which filled up the beds, weeds were removed, and that was all. Two kinds were grown: the sweet violet and the Parma violet. As regards the latter I should say that at Grasse it had been widely grown for a long time. What was its origin? No one knows. What can be affirmed is that it was spoken of in this region from 1755. Moreover it was not the beautiful variety of Parma violet that we grow today, but a small and rather pale Parma violet. My father still grew it when he was young, and I can remember it perfectly. Besides, some old growers from around Grasse have kept it; the younger ones grow sub-varieties that are more beautiful and more double (1). The fact remains that during the first half of our century this cultivation, as far as it concerned the perfumery business, was the object of considerable enterprise, and with the orange trees, a source of wealth. Prices varied from 3 to 12 francs the kilo (of flowers without the stem).

(1) With regard to these sub-varieties, I must establish certain facts, although they imply almost the opposite to what I suggested

earlier, when saying that Parma violets never, or hardly ever, produced seed. It is true that I was speaking of the present time and of our beautiful Parma violets in the climate of Paris. Well, those of the South of France which we are considering now were not well grown or cared for. At the end of each season the flowers were very small and became less double. In this state not much more is required for seed to be produced, and I have no difficulty in believing it, since the double white and double rose violets behave in the same way in the climate of Paris. Thus one could well believe in the creation of these beautiful sub-varieties such as 'Parme de Toulouse', 'Parme ordinaire', 'Parme sans filets'.

I have cross-examined elderly growers to try and trace the origin of these fine sub-varieties. Some told me: "They came from Grasse just as we know them, having sprung up spontaneously." Others on the contrary alleged that it was by selecting from the stolons that they had obtained these improved kinds. In any case it was not easy to establish precise facts on this subject. I am not surprised at the indifference of these growers; those in the vicinity of Paris are equally indifferent with regard to new kinds. If they have a higher market value, they take note of them, recording the fact and not bothering themselves about either the origin or the raiser. Finally, whether from the soil and the climate causing seed to be obtained, or from improvement of the stolons by cultivation, it can well be granted that the beautiful sub-varieties of Parma violet were born in those regions.

The Second Empire[64] and the railways came to change the aspect of this trade at a stroke. The fashionable habit of spending the winter in the South of France created an outlet for this flower in the towns of Provence. Nice[65] becoming part of France again also greatly promoted the sale of violets. In the winter months the price of Parma violets rose to 40 francs the kilo, which was still relatively modest compared with prices reached a few years later.

In his works Alphonse Karr[66] sang the praises of the flowers of the Midi. An enthusiastic amateur of roses and violets, he may be said to have been the instigator of consignments of fresh flowers. The stimulus having been given, gardens of flowers were created along the whole coastline. Fields of violets did extremely well there and were an immediate source of wealth for their owners. The destination of the flowers was changed; instead of going to the manufacture of perfume they were dispatched in almost all directions. It was a general frenzy. Impossible prices were known (100 francs and more the kilo with stems, especially during harsh winters when there was a shortage of violets from the Paris area on the metropolitan market). The manufacturers of essential oils and scent now had at their disposal only rejects and the flowers of the end of the season. Consequently the growing of these flowers was wildly extravagant. From Nice to Toulon there was nothing but fields of violets. It was the high point of the violet of the Mediterranean coastline and the death of the open air cultivation of violets in the vicinity of Paris.

I shall explain the reason for this when I speak of the latter cultivation.

These fine times could not go on for ever (1). Alphonse Karr, as I have already said, and after him M.de Solignac[67], created and caused to be created the cultivation of many flowers: roses, carnations, stocks, anemones, mignonette etc. came themselves to share in the glory of their sister. Tastes fluctuated, violets had to share the favour of the world of fashion with other flowers of the time. Prices were affected by this and became normal again. Parcel post favoured consignments, cultivation split up and extended towards le Var[68]. One group, the older growers, continued with the old varieties for perfumery, whilst the others, specialists in cut flowers, were already towards 1875-1876 searching for varieties with more beautiful flowers.

(1) It is true that from 1855-1856 up to 1874-1875 is a fairly good reign, the more so since at this time the varieties of violets were not numerous. In the Midi the Parma violet formed the basis, then the sweet-scented quatre-saisons or improved quatre-saisons with large petals, and lastly 'Wilson'[69] (or 'Violette de Constantinople').

For a long time they favoured 'Wilson' (a large violet with a long stem and big petals, but having less scent than the sweet-scented quatre-saisons). Its faults were to be a little pale in spring, and having very thin flower stems that scarcely support the large flower.

Then came 'Czar'[70] (popularly known as 'The Russian');

and an improvement of 'Czar', which was named 'Reine Victoria' (1), or 'Czar bleu'. Its petals were a little longer than those of the 'Czar', but it came into bloom slightly later.

(1) This 'Reine Victoria' must not be confused with 'Luxonne', which in many places was also called 'Reine Victoria'[71].

Finally towards 1886-1887 there appeared in my seedbeds, almost spontaneously, a violet with large flowers, long petals, very long flower stems; in brief a magnificent variety that by chance came to be called 'Luxonne' (2). This was followed by a very beautiful improvement of 'Luxonne' that its raiser named 'Mme.E.Arène'[72]. It was a great success, but the name was not widely known. Although much more beautiful, it was too similar to 'Luxonne', and in the trade at the markets they were confused. Only amateurs were able to appreciate the advantages and the difference.

(2) I was teased a little about this name 'Luxonne' in 1888 at the concours général[73], where I had presented a fine show of this violet. Eminent botanists said to me: "Where then did you get this name?" I replied that I would have much wished to change it, but that it was unfortunately not possible, it was too well known to the public. One of my clients at Solliès-Pont[74] had even obtained an improvement of this kind which he called 'Mme.E.Arène', but it led to a muddle; commonly it was known as 'la grande Luxonne'. 'Luxonne' had been christened in the markets, and the name was adopted.

At the same time several specialists in the Solliès-Pont area grew two of my fine large-flowered varieties: 'Souvenir de Millet père' and 'Gloire de Bourg-la-Reine'. I have to say that they were only moderately successful. The first was too delicate in the summer season and only grew there with difficulty. 'Gloire de Bourg-la-Reine' grew perfectly well there, but did not travel satisfactorily, because its petals were too compact and arrived all crushed. Finally, to end this rapid historical account of violets I come to one of my sweet-scented varieties which seems to be excellent, especially for the South of France. It has single flowers, is vigorous and very pretty; so far it also seems to be very suitable for dispatch to the markets. I refer to 'Princesse de Galles'. To sum up, four varieties form the basis of violets grown for consignment to the markets. They are: 'Quatre-saisons odorante', 'Czar', 'Luxonne' and 'Princesse de Galles'.

The manufacturers of perfume continue to use the old varieties, which contain much more essential oil for distillation.

Violets in Haute-Garonne[75], and a few words on Angoulême[76]

Before leaving the violets of the Midi I should draw attention to a region that supplies the markets of Paris. For a very long time now Parma violets have been grown in the neighbourhood of Toulouse, especially in the villages of Lalande[77] and Aucamville[78]. Parma violets produce very fine results here, doubtless because of the local climate or selection of cultivar. It grows splendidly, the leaves reaching 20 to 25 centimetres; the flowers are large, with very firm petals, and they are borne on stems that regularly attain a height of 20 centimetres, which renders them suitable for making up all kinds of bunches and bouquets.

The violets are cultivated in a very summary fashion in large beds that are not even renewed each year; everything is left in the open air. For the last few years, since this cultivation has brought in a fairly good profit, many growers have been giving their crops protection under glass, and an important trade is developing year by year.

In view of the beauty of its flowers, the foreign traveller is pleased to buy Parma violets in autumn and spring, and the flowers are dispatched to nearly all places. As a result of this there has been a great increase of cultivation and a rapid development of consignments of violets to the markets.

And if this marketing is not always on a very large scale it must be attributed to the winters, which are sometimes quite severe at Toulouse, and cause dispatch to be delayed and, because of this, reduce the importance of the trade.

Quatre-saisons, as well as to a small extent 'Czar', are also grown, but their cultivation has not increased, and they remain a product to satisfy local demand.

Finally I must mention another town whose Parma violets were much talked of some fifteen years ago. Many florists spoke of the beautiful violets of 'Parme d'Angoulême'. They were even announced in several catalogues, saying that the flowers reached the size of a five franc piece[79]. As I was anxious to ascertain the truth about this, I sent for some plants and grew them. I was soon disillusioned; they were exactly the same varieties as ours, and what is more the flowers were not so pretty. In order to be quite certain, when I was passing through the region I visited several places where Parma violets were grown. I must say that they were very well grown in frames and heated during the winter, in a similar way to that used by us in forcing violets in Paris. Results were exactly the same; and as the cost of production is high it has not been possible to establish consignments elsewhere, and sales have remained purely local.

Since the cultivation of Parma violets developed in the South of France before it did in the neighbourhood of Paris, I have been obliged to pursue the course of events up to the present time, and this has taken me away from my subject. I am compelled to go back in time, and I ask my readers to pardon me for this.

At all events it has been established that two styles of cultivation developed, in places 1,000 km. apart, each following its own requirements and climate and, what is remarkable, almost at the same time, that is to say about the middle of the eighteenth century.

1700 to 1750
Violet cultivation in the Paris area; how it began

As I said earlier on, it was about the beginning of the eighteenth century that violets came to be sold commercially, but they were not then cultivated. They were gathered in the woods surrounding Paris, such as the Bois de Verrières, de Clamart, de Meudon, de Boulogne, de Vincennes, and further out the Forêt de Bondy, the Bois de Romainville etc.

The same class of people who carried on this little trade about two hundred years ago still do so today. They were, and still are, gatherers of medicinal herbs. Each season provides them with a plant: periwinkle, wild radish[80], mint, heartsease, valerian, violet etc.

These people do not even give themselves the bother of making the flowers up into bunches. They sell them in small quantities, wrapped in cloth the size of a large handkerchief. They pick them in various colours as they find them: blue, pale colours or white. Many of these parcels, especially towards the end of the season, are sold for enfleurage[81] or for drying. The dealers of the markets made up little bunches from these parcels of violets and disposed of them on the streets of Paris. This is how the trade in violets began. The contacts between those who went out gathering violets and our market-gardeners in the Paris area, who saw these little bunches of wild violets being sold nearby, aroused in the latter a desire to grow violets.

Indeed, about 1755 several beds were tentatively planted in market-gardens, the plants being obtained from hedgerows. They were planted in a mixture, without distinction of varieties. The bed was left for several years in the same place, to the detriment of the violets. They were small, rather pale, similar to the wild ones. However, the sellers noticed that bunches of uniform colour sold best; there was most demand for violets of dark colour.

1750 to 1780

Time passed; thirty years elapsed, leading on from this initial cultivation of violets, and they began to make a selection as to pale violets and blue ones, as well as white. The earliest flowering varieties were chosen, those which bloomed in autumn and in spring, from which they were called quatre-saisons violets.

This small-scale cultivation was established before 1780 in many of the areas around Paris: at Vincennes, Charonne, Bagnolet, Saint-Cloud, Massy-Palaiseau, Fresnes-les-Rungis; but not all these villages continued with it. Only at the village of Fresnes-les-Rungis[82] was the cultivation extended, and there it was better managed.

Either by chance or skill they obtained and fixed a good uniform strain of quatre-saisons, which however produced very few flowers in autumn. The plants were frequently renewed, and yielded fine flowers.

How violets were sold in Paris in 1780 and the following years

Unfortunately the troubled times of this era were not very encouraging for the cultivation of flowers, and it was difficult to sell them; doubly difficult in that they were sold directly by the growers, that is to say by their wives.

In 1860 I knew a fine old lady aged 85, still very able-bodied, who used to tell us about the troubles in Paris during these years: "I was quite a young girl at this time", she would say, "and I used to sell the violets that my father grew. We would take up our positions in the best districts of Paris, having in front of us a small éventaire[83] (this was a kind of flat basket with a board as its base, and attached on one side to one's belt, a cord from the other side passing round one's neck to keep the tray horizontal). The bunches that were all ready were placed at the front and, with both hands free, we made up the little bunches as we walked along. We stopped mainly at the corners of the streets where there were the most people. At the beginning of the First Empire[84]", she added, "they didn't want to let us go on selling like that, you were supposed to put in for a permit and get a place. But damn that, we sold them all the same, and ran away when the police arrived."

And so we see from this little note that, even after 1795, the trade in violets was not on the vast scale that it is today.

Meanwhile the Parma violet was grown in private gardens. It was kept in frames at the base of walls, but no one thought of growing it for the markets.

How had it been introduced into well-to-do houses? It is impossible to answer this question. I remember very well that, when I was ten years old, about 1855, my father was talking to a very old gardener who grew the Parma violet and asked him how it had come to be grown. His reply was that he had always known it, and that it was handed on from gardener to gardener. This man, from his great age, might very well have been able to remember 1770.

But I must leave the Parma violet in order to continue with the development of the single violet, the trade in which had been launched.

1795 to 1825
Cultivation draws nearer to Paris, and spreads

I said then that about 1795 the village of Fresnes-les-Rungis was about the only place to carry on the trade in violets. However, at Montreuil walls were built for growing peaches on, and in front of the espaliers[85], in borders, violets were planted, and they were sold together with small white hyacinths. But this cultivation did not expand, and only offered the peach growers a slight addition to their profits. For almost twenty years now it has practically disappeared.

To the south of Paris things were different. From Fresnes-les-Rungis cultivation moved on to Châtenay, Bourg-la-Reine, and more especially to Fontenay-aux-Roses. There everything was just right. Specialised rose growing left land free. In order to rest it they planted violets, other flowers being of little account[86].

So from 1795 to 1825, that is to say in the space of thirty years, the trade in cut flowers increased; violets and roses became commercial commodities.

1825 to 1835
Improvement and considerable extension of cultivation

Although France was not at this time endowed with a stable government, there was a feeling in the air of the need to live in tranquillity, to go and plant one's cabbages, to use the colloquial expression. Holdings were reorganised, there were early crops and forced flowers. In town parties and receptions had an effect on trade: people wanted to decorate their tables and drawing-rooms.

To satisfy this demand firms were set up, the number of places growing violets increased and expanded. Then came the reign of Louis-Philippe[87] to speed up the pace of development; receptions were numerous, flowers were required.

In winter the violet could well have been just what was needed. Unfortunately there was only the Parma violet, that a few cultivators grew with the help of heat, beginning to force the flowers from the autumn; but it could not provide sufficient flowers. Between 1835 and 1836 a gardener-nurseryman from Fontenay-aux-Roses, Jean Chevillon[88], obtained a quatre-saisons violet which was almost perfect, that is to say fragrant and as beautiful as the ideal type of the woods, and yet decidedly remontant. Picking could begin in September, and when grown in frames or in a very sheltered position, flowers were picked in winter; and so it created a sensation. It seemed that the gardener-florists who grew this kind of plant were just waiting for this result. It was sold at a relatively very high price, small stolons or runners fetched one franc apiece. Despite this price, and precisely because of it, it was rapidly propagated. Certain growers, overestimating, carried the development of this cultivation to extremes. Among others at Bourg-la-Reine, a man named Vogt[89] brought the number of his frames up to 1,800, both Parma violets and quatre-saisons.

1835 to 1843
Trading is regularised; violets sold wholesale

About 1838 at the village of Fontenay-aux-Roses all the fields that were not being used for roses were covered in violets.

As dealing in and selling violets was carried on easily (1), greater attention was paid to the plants, and this brought about the appearance towards 1840 of another sub-variety with a larger flower under the name of 'grosse bleue' (also called 'Ravageot'[90]). This late-flowering variety, with dark violet flowers, only bloomed the once in spring, and followed after the quatre-saisons grown in the open.

(1) I omitted to say that from the middle of the First Empire trade in flowers by the growers had been regularised. Sale was carried on wholesale in the rue aux Fers around the Marché des Innocents and the famous restaurants, fashionable at the time, Baratte and Bordier[91]. Maison Baratte is still there, but the name of the street has changed. The Marché des Innocents was closed down in 1855, and from this date the flower market has followed other products to the Halles of the present time.

Commercial growing of Parma violets, undertaken at first about 1830 only tentatively, became quite extensive. In the reign of Louis-Philippe it passed from private houses into commerce. To begin with only one variety was known, fairly pale blue-mauve. Then some time about the years 1835-1840 appeared a variety without runners[92], a little darker blue in colour, very double. Unlike its precursor, it never bears seed.

This variety was very pretty, more compact, and had a darker flower than its congener. It was soon in the hands of specialists, and I have not yet been able to establish its origin with certainty.

The first time my father saw it was at the holding of a grower at Bourg-la-Reine called Paré[93]. He had found it by chance among his Parma violets. Having separated it from the other plants, he propagated it and cultivated it specially, thinking that he had obtained it by chance and that it had first appeared on his land. My father did not share this view of things, the more so since it appeared in different places at the same time. M.Mascré[94], of Sceaux, had received it the same year from well-to-do houses; and since then this firm has grown it profitably for more than forty years. It is evident from all this that in 1848 in the Paris area there were still only four varieties of violets grown for cut flowers: two Parma violets and two single blue violets. There had, it is true, been tentative attempts at growing double blue violets, but they had not been a success.(1)

(1) This very pretty variety is suitable for gardens, especially for borders, and flowers profusely. It is quite fragrant, but has never found favour as a cut flower. For more than seventy years from time to time attempts have been made to launch it, but they have always failed, as it is not thought pleasing in bunches.

1843 to 1859
Violet cultivation, especially in the open, is all the rage

During a period of ten years no new variety appeared to create a diversion, but the trade in violets had greatly increased.

The Republic gave way to the Empire[95], there was a breath of unparalleled luxury in France; there was a demand for flowers at any price. Bunches of violets fetched fabulous prices; so the growers had a thoroughly good time. Violets were planted in the open, on hotbeds, forced in glass-houses, finally everywhere where one could gather violets at all times.

From Fontenay and Bourg-la-Reine planting extended to Sceaux, Châtenay, Verrières-le-Buisson. Hundreds of hectares of land were cultivated, and it is a remarkable fact that while this plant was spreading around Paris it was leaving its birthplace: at Fresnes-les-Rungis cultivation was abandoned, and there is no longer any trace of it there now.

At this period cultivation of flowers was most remunerative. I have seen ordinary growers of unprotected plants, in the autumn season, sell 300 to 400 francs' worth of violets in a single day at the market. However, exorbitant prices had not yet been reached.

1859 to 1870
Two excellent varieties make their appearance; prices obtained

I have chosen these twelve years because it was during this time that several novelties, and in particular two sensational novelties, made their appearance.

One of them was imported from Turkey to the Château de Segrez by the president of the Horticultural Society, M.de Lavallée[96]. It was while he was travelling in Turkey that M.de Lavallée came across it in the mountains.

However, he did not bring it into this country until about 1871, but it had already been grown in the South of France for a very long time, even by those growing for the perfume manufactury. I had received it from the Midi in order to test its suitability for forcing. It was like the one M.de Lavallée had given me. It was, and still is, a violet with large, long petals which form a slender, pale violet flower of good size, held on a long stem that is not very rigid. It was sent to Paris and sold under the name of 'Wilson'. It was this violet which, pollinated by the 'Czar', produced the beautiful variety 'Luxonne'. This put an end to the use of 'Wilson' in the cut flower trade.

The other was popularised by one of the biggest growers of Verrières-le-Buisson[97]. He set to work growing and rapidly multiplying a kind of quatre-saisons violet, which did not at first reveal the future that lay before it. A very strong peduncle; petals that were more compact, bigger, more erect, of a fine colour; a very good scent: such were the qualities of this quatre-saisons, which was much easier to make up into bunches, an important question with regard to labour-saving.[98] It immediately superseded the two quatre-saisons (1) that were grown. The fields were covered with it, and the gardens of those who engaged in the forcing of violets were filled with it.

(1) I speak of two quatre-saisons because for many a year the growers of violets from seed had by selection formed two sub-varieties. Some thought good to choose the paler flower, which certainly bloomed earlier in the winter, but became defective at the end of the season, because it was too pale. The others had selected the blue violet, with a darker, more purple flower, which had more sale at the end of the season but produced less in the winter. However that may be, this formed two quite

different types and, a curious effect of nature, the type with dark flowers produced a lot of seeds, whilst that with pale flowers produced hardly any. I cannot very well explain this curious fact; however, I have ascertained that in general the earlier plants flower, the fewer seeds they produce. The 'Quatre-saisons Semprez', which is the finest of all quatre-saisons, hardly ever sets seed, and if it does, it is degenerating.

In a word it was the only violet known on the Paris market, bunches of it were always sold at a higher price than those of other varieties. The market itself named it; when anyone wished to speak of it, or let its higher value be known, they simply said: "It's the Semprez". This was the name of the grower who had launched it into commerce. I say launched deliberately; this grower had not obtained it from seed, he had it from Vilmorin's[99], who gave him three or four varieties to try out, without attaching any importance to them.

This violet came to the fore as a result of careful work, and this is how an exceptional variety was established without any recommendation, solely through its fine qualities.

It has lasted for thirty-five years, and despite new varieties that make a lot of noise, it seems unlikely that it will soon be replaced, especially for forcing.

While speaking of forcing, I should say a word in passing about the prices fetched by forced violets and those grown in the open from 1860 to 1870. These prices varied a lot according to the season. Violets were sold, as they still are today, from September up to the end of April.

To give a clear impression of prices at sale, I must explain what the bunches sold wholesale at this time were like. They were usually made up in the evening, and even far into the night during the busy season, on the premises of growers who cultivated violets on a large scale. It was not unusual to see fifteen to twenty people around tables making up bunches.

A word about how these bunches were put together. Usually two handfuls of violets were gathered up, a tie drew them together and secured them, then they passed on to another operation. Others arranged leaves close together all round the bunches so as to form a green collar, which made the bunches very attractive.

These bunches, which were almost spherical, could be from 8 to 10 centimetres in diameter and contained 250 to 300 violets, depending on the size and beauty of the flowers. As with all articles of luxury, the higher the price of the flowers the smaller the bunches became.

As I said, prices varied a great deal. From the autumn season they fetched 0fr.35 to 1fr.25; during the winter months, between 1fr.50 and 3 francs. In the Christmas and New Year holidays prices rose to 4 and 5 francs, according to the mildness or severity of the winters.

Prices of Parma violets always remained a little higher. As they were not grown in the open, there was no confusion. The price varied according to the season from 0fr.75, to 8 francs for the winter months. The bunches were different; they were – and still are to this day – much more finished; that is to say the violets were individually arranged around a handful of straw so as to form them into a flat bouquet with every flower visible. These bunches were between 12 and 16 centimetres in diameter, and they too had a collar of leaves, often supported by periwinkles. This has always produced pretty bunches, and so they were much appreciated for evening receptions and balls. When someone walked by with one of these bunches, it left behind a most sweet fragrance.

1870 to 1880
An avalanche of novelties: 'Czar', 'Millet père','Gloire de Bourg-la-Reine'

With these varieties we approach the year 1870[100]. A single violet with large flowers had been noticed in both regions; but the misfortunes of France temporarily put a stop to horticultural work, and it was only towards the end of 1871 that one heard of new varieties of violets. The 'Czar' was one of them.

My father had obtained an enormous violet, with very large flowers of a delicate blue-violet colour, large leaves, deliciously fragrant, erect, flowering very early, but fairly tender, and above all having the quality of giving flowers in December and January, when properly treated to do so.

We sold direct to several Paris florists at a very high price. The flowers in these bunches were raised to form a flat surface, like those of Parma violets. These were the first single violets to be made up in this manner, and began this special method of bunching violets grown in frames. This variety was the starting point for the large-flowered violets; however, it did not have the priority in commerce owing to the fact that we did not put it into circulation until 1876. During this period the 'Czar' made its appearance. It was obtained by Thomas Ware[101], of England, and introduced into France by Lemoine[102] of Nancy. Yvon[103] of Paris had it from both of them, and showed several plants at various societies, notably at Paris.

Its entry into the world of cut flowers was a most difficult one, although it was a plant of good quality. To begin with it is very hardy, even in the open, has very fine dark green leaves, held up on robust stems. The flowers are of moderate size; the petals, long, thick and strong, form a compact flower of a dark violet; its fragrance is very sweet; the peduncles are 15 to 20 centimetres long. In a word, at that time all these qualities made it a kind of violet that was, though not perfect, one of outstanding merit.

Some amateurs bought it, and we grew it on a large scale for the Paris market. But there, as with many new things, we came up against routine; these pretty violets were not wanted, they were too big, the stems broke when bunches were being made up. In brief, it took more than four years of perseverance to get it finally admitted into the cut flower market.

In gardens its position as a new plant was different. It was quickly realised that it was hardy, formed very pretty borders, and provided beautiful flowers in the spring. In winter cultivation it took second place after the 'Semprez' violet, its fault being that it did not produce many flowers in the winter months. Strangely, despite the catalogues, which gave its name, that is to say 'Czar', the markets called it simply 'The Russian', the name by which it is best known.

In these same years the need was felt for new varieties; all the specialist growers were on the lookout. An old Parma violet was relaunched. It was very pretty, had dark blue flowers with a small red petal in the middle. It was known under the name of 'Marguerite de Savoie'[104] or 'Marie-Louise'[105]. The latter name has prevailed. Its origin is far from certain. Old gardeners have told me that it dates back to the First Empire and that it was raised at Malmaison[106]. I am prepared to believe it; yet it is not mentioned in any book of that time. You have to look in works of 1821[107] to find the first reference to Parma violets. However that may be, it is a beautiful variety that comes in succession to the first ones, especially in spring, when the latter become pale. It retains a lovely dark blue colour and makes splendid bunches. However, it does not manage to gain a place as a commercial violet.(1) As it does not bloom sufficiently in winter, it is rather a plant of the autumn and spring. It has remained a plant for the amateur.

> (1) One of the reasons[108] why it is not generally accepted as a cut flower is that in the middle of the blossom is a small red petal which, though it does not mar the bouquet, makes it very novel: this is not to everyone's liking.

And then it faced serious competition from the fact that almost all the growers of Parma violets, in both the Paris zone and the Midi, made selections from their plants, ever sorting out those with the darkest flowers, the most compact, the biggest. Having done this, and with the aid of skilful cultivation, they obtained a sub-variety no longer having anything in common with the old Parma violet of former times.

A conclusive instance of this was that, when I found Parma violets in the gardens of old houses, you would think there were two quite distinct varieties. This improved form has not changed its name. However, it comes to us from the Midi with the name of 'Parme de Toulouse'[109], but after having been grown for two or three years in Paris it is no longer distinguishable from the other forms.

The notion of new varieties made great progress. We raised a single violet, rose-blue, very early flowering, blooming well in winter. At the same time it is a novel kind and flowers generously. But all varieties that deviate from the colour violet have difficulty in making their way and it did not please as a cut flower. It was named 'Lilas', and is still grown in collections.

Under the name of 'Belle de Châtenay' M.Paillet[110] distributed a double white variety, with very large flowers lightly tinged with blue that are not held up firmly by their stems. It comes into flower very late, and is a violet for the spring. It had only a moderate success. The flower colour is variable: in cold weather the white of the petals is greenish and the blue not well-marked; in warm weather the colours are more distinct, and the mixture of white and blue is not displeasing. In short, it is a good plant for the amateur.

In this same year a single white variety called 'White Czar' came to us from across the Channel. It is very freeflowering, pure white, with long mat leaves of pale green. Its weakness is that the flowers are not compact; the petals are thin and frail. It produces far too many runners, which on occasion exhausts the parent plant.

For nearly twenty years I have kept on sowing this plant, but have not succeeded in getting a sturdier form. When grown in the open it suffers during harsh winters. It is easily increased from seed and flowers very freely.

In 1875 'Brune de Bourg-la-Reine'[111], one of the most beautiful violets, was put on the market. In bunches and in a bright light its colour takes on various changeable tints that are quite exquisite. It was obtained from a cross between 'Czar' and 'Lilas'. It has the form of 'Czar'; the petals are, however, slightly more elongated, the peduncle tall and firm, holding the strongly scented flower upright. Its colour is bluish-purple, and the bunches emit metallic reflections with the prettiest effect. Its principal flowering is in spring, but it yields quite well in autumn. The leaves are a delicate green, long and held very erect.

In the same year, 1875, from sowings of 'Czar' I obtained a sub-variety with petals that are quite rounded and very firm, darker in colour than 'Czar'. The scent is just as pronounced but more delicate; the leaves are more deeply serrated[112]; the peduncles, leaves and flowers much stronger, more erect. It is a plant of two seasons; it is in bloom early in the autumn and late in the spring. I called it 'Czar bleu', or 'Reine Victoria'. It received the second name the following year, as it appeared in various places at the same time, and under this name of 'Reine Victoria'(1). So as to avoid confusion I left it with the two names, which were already known. It was precisely the same plant.

(1) It had been named in the Midi in honour of Queen Victoria, who stays there each year.

At last, in 1879, I had the good fortune to obtain a violet with an enormous flower, almost the diameter of a large pansy.

This novelty, which was to be the ancestor of the pretty flowers we have today, was superb, with wide, strong, rounded petals that opened very well to form a broad flower – blue, delicately tinted violet – with a subtle, sweet fragrance. The peduncles were erect, 15 to 20 centimetres long; the large leaves, 6 to 9 centimetres wide, of a good dark green, were markedly denticulate and held very upright.

All these qualities together made it a plant of the first rank. Very hardy, it stands up to our severe winters as well as do the quatre-saisons. It was named 'Gloire de Bourg-la-Reine'. It is very good for forcing. Like many of its predecessors, it is more a violet for the two seasons, autumn and spring, than for winter.

By a freakish effect of sowing, when I obtained this variety I found another plant, their roots tangled together. This was the pretty little violet 'Armandine Millet'[113], whose green leaves have a border of ivory white.

In the autumn the flowers are a fine rose[114]. It blooms continuously and is as hardy as any violet. It is pretty in all positions and at all times of the year: in a border, planted out in a mosaic bed[115], even in grass. Provided that it is sheltered it flowers well throughout the winter. It is one of those rare plants that have leaves variegated green and white, and blue flowers.[116]

The last period, 1880 to 1896: fine new varieties

These last fifteen years, although they have not given us a large number of varieties, have brought us some beautiful violets. I leave out on purpose those varieties which, though pretty, do not possess any individual character.

About 1881 M.Paillet put on the market 'Viola odorata rubra'. It is a magnificent little variety with bright red flowers, often a rosy red, especially among seedlings. Its bearing and foliage are reminiscent of our wild violets, although the leaves are darker and flatter in appearance. It has an exquisite scent. Its only defect is that it hardly flowers at all in the autumn; but it makes amends in the spring, forming a garnet-red carpet that is wonderfully beautiful. Also, when planted under trees or shrubs it is very choice and recherché.

About 1883 in the South of France the violet 'Wilson', pollinated by 'Czar', produced a very pretty violet. As it appeared at the same time in several places where the two varieties were grown, no particular person can be said to have raised it. It was given its name by chance at the Central Markets.

It was dispatched under the name of 'Grosse Wilson', or 'Wilson Extra'. But those who were irritated by the name 'Wilson' called it 'Luchon', and then 'Luxonne'[117]. It developed rapidly, and being distinct the name 'Luxonne' prevailed and became firmly established. It was, and still is, one of the largest flowers, with very long and slightly thin petals. It makes pretty bunches. It flowers very freely in autumn and in spring. It was a fine acquisition for the Midi, since in Paris growing it in frames causes etiolation, the flower and its peduncle become thinner (1), and it does not give the results one was entitled to expect of it. Although it does have a scent, it is a little lacking in fragrance.

(1) It is the same with all large-flowered violets. In order to get very beautiful flowers in winter in the climate of Paris they must be kept in cold frames, and only very few good clumps planted in each frame.

Several sub-varieties showing some improvement came from its seed, but they all had the same type of flower.

I said elsewhere that, especially in the climate of Paris, it is extremely unusual to find seed on Parma violets. This is strictly accurate: it is in a way a phenomenon.

However, this phenomenon occurred to my advantage. One of my friends[118], a great grower of Parma violets for cut flowers, made a careful selection of Parma violets in order to obtain beautiful flowers. He did this with passionate energy, wherever he found Parma violets growing, especially where they were no longer tended and almost like wild flowers, never having been given protection. The further they seemed to be from careful cultivation the better he liked them. And he was rewarded for his efforts, and soon gained a high reputation in the cut flower market.

But on the subject that concerns us, the most inexplicable event occurred: on the selected variety, with double flowers of increased size, he obtained several seed capsules.

I urged him to record the fact at the Paris Horticultural Society, and he did so.

While this was going on I bought some of his fine plants, and as I knew he had gathered some seed, we were able, as a result of unceasing close attention, to discover four capsules on more than 50,000 plants. This was not much, certainly, but it was sufficient for us to obtain a very beautiful variety with rose-coloured flowers.

This Parma violet, dedicated to Madame Millet, was for me a valuable acquisition. It was a new departure for Parma violets as regards colour, since previously there had only been the one.

It is a bright rose colour, slightly mauve, the flowers always very double and full, with numerous petals which form little roses when it is in full bloom. It has the same scent as the Parma, but a little more pronounced. It blooms to excess from September right up to May, and carries the flowers very erect. The leaves are the same shape and consistency as the 'Parme ordinaire', a little shinier and glossier but slightly smaller. In cultivation it behaves exactly as does the Parma violet.

1884 to 1886 witnessed the appearance of 'l'Inépuisable'[119], a cross between 'Czar' and the violet 'Semprez'. It has very long petals and is very fragrant. It flowers to excess from September until April, but becomes a rather pale violet from the effect of the sun.

Then came 'Bleue de Fontenay', a beautiful little variety that blooms very generously in spring: very deep violet, dark green leaves. It was obtained by a grower at Fontenay from repeated sowings of 'Quatre-saisons bleue'.

This variety does very well in borders, as it sends out few runners. Its flower, though not competing with our large violets, is admirably suited to making up simple bunches.

Towards the end of 1886 I received from England a white Parma violet under the name of 'Swanley White'[120]. On the other hand it was announced as having come from America with the name of 'Brazza White': it was also grown at Bayonne under this name. I collected together and bought all these different violets and became certain that they were the same plant. I only regret that I cannot give the name of the person who raised it.

As soon as it appeared this variety made a great stir among amateurs of violets. We already had the rose-coloured Parma, now there was the white. There was an endeavour to obtain a seedling of a distinct colour, but unfortunately the lack of seed always held us back. However that may be, this variety is beautiful.

When it grows well, in full vigour, its flowers are like little gardenias. As with all Parmas it flowers abundantly, but is a little less prolific in December and January. It has fine large leaves, perhaps the largest of this race, and it grows and spreads as vigorously as any other variety. It has slightly less fragrance than others of this race, and a slight defect, which is that in the fine sunlit days of spring its petals take on a bluish tint, and this undoubtedly recalls its forebears.

Some years ago it came into the sphere of the cut flower trade, rather for consignment abroad than for sale in Paris, which is still faithful to its old customs and likes only the colour violet.

However, several cultivators grow it in pretty large quantity, but with the intention of selling them as potted plants in spring. And yet people reproach it because its flowers do not stand up well; the flower is laxly supported and remains among the leaves.

I said just now that its petals become tinted at the tips with a bluish colour, the colour of 'Parme ordinaire'. I am personally convinced that this is due to the nature of its origin.

It has been reported to me that it was obtained from seed[121]. I do not believe this myself, and for the following reason. Since I first grew this plant, practically every year without exception I have had one or two plants, even more, that had three or four crowns fairly distant from the main one, and out of four crowns two had white flowers like the variety itself, and the other two had flowers of 'Parme ordinaire'. Now I have always heard it said and recorded that plants obtained as sports frequently give examples of reversion to the original plant. As it often produced these unwonted cases, I concluded that 'Swanley White' was also the result of mutation[122].(1)

(1) This occurrence seemed so strange to me that I took several plants to the Horticultural Society to have it recorded, for this observation seemed to me to be a rare case with violets. I had never seen anything like it among ordinary violets – double white, double blue and double rose – which have been cultivated for centuries.

I forgot to say that, from its colour being white, this variety is quite suitable for forcing.

In 1887 to 1888 the firm Forgeot[123] put out a very dark quatre-saisons violet of the original type but distinguished by a quaint anomaly. In spring each year, when the sap rises, all the veins of the leaves become strongly gilded, and this causes a variegation as unusual as it is pretty. This striping of the leaves lasts for about three months; after this time the effect is not so attractive. Its flowering, especially in spring, is of the finest. It gives a profusion of medium-sized dark violet flowers, which make a very good effect in a border or mosaic flower-bed.

Concerning the influence of resin on the colouring of leaves[124]

Regarding variegated leaves, I must mention in passing an interesting experience I had a few years ago.

I was visiting an estate at Épinay[125], and when I arrived at the vast park which is part of it, I was struck by an immense carpet of scentless violets, whose leaves of a most beautiful golden yellow with green veins were ravishing.

You can imagine, dear readers, whether I, a great lover of violets, was happy! Here, I told myself, is a variety which, although it has no scent, will enrich my collection and be ideal for planting in borders in woodland and elsewhere. And so without wasting time I asked for permission to take up a few plants in order to multiply them rapidly, and this I did with great care.

But alas! what was my disenchantment when, having carefully planted it in good open soil, I saw it produce completely green leaves and remain so all year long. I then recognised it as a variety I had very often come across in the woods, especially in mountainous places, and at the same time I accounted for the phenomenon of the gilding of the leaves. It reminded me that the woodland area where I had taken it consisted entirely of firs, which renewed their leaves each season, dropping a considerable part of them on the soil, and by so doing forming a good thick mulch, composed almost exclusively of a basis of resin, which had caused the abnormal change of the leaves of the violets.

I am happy to be able to say also that the fact I have drawn attention to has been recorded by one of our leading horticulturists.

M.Maxime Cornu[126], director of glass-houses and gardens at the Muséum de Paris, said to me one day point-blank: "Monsieur Millet, some time ago I thought I had found you a beautiful variety of violet with a fine golden-yellow variegation of the leaves; but I have to tell you that I was straight away undeceived. As we had taken these violets from under large fir trees and they were replanted in ordinary soil to their great detriment, they became green again. From this I concluded that the resin alone caused the colouring, or I should rather say discoloration, of the leaves".

And so there was no doubt about it, the case I had observed was confirmed two years later by M.Cornu, without however any harm to the growth of the plants having been recorded. Is there not something instructive here with regard to the colouring of leaves?

This example leads me away from my subject. Returning to 1889, I must mention in passing, although it be only an improvement of 'Czar bleu' or 'Reine Victoria', a variety introduced by Bruant[127] of Poitiers under the name 'Wellsiana'[128]. Along with 'Gloire de Bourg-la-Reine' it is one of the sturdiest and most compact varieties.

With well delineated rounded petals of a fine deep violet and with metallic reflections, it is supported on stiff dark peduncles. The flower is suitable for making up bunches or for planting in pots for the markets. It does not bloom much during the winter, but has the advantage of giving a good yield of early flowers in the autumn, and a second crop in all its splendour in the spring.

A few notes on 'Princesse de Galles'

Approaching the year in which I am writing these notes, I reach 1891, when this beautiful variety 'Princesse de Galles', which is all the rage at present, first appeared. Many people remain ignorant of its origin. As a matter of fact it was born – I should say appeared – among the violets of several growers in the Midi at the same time.

It was born at Bourg-la-Reine in 1889. In February 1891 I exhibited a few plants in pots at the concours général, where it was considered to be an admirable variety.

Our well-known explorer, M.Dybowski[129], who was in charge of the horticultural section at that time, asked me for a pot. Although I had very few of this variety, I did not dare refuse. On this occasion he even observed that I ought to name it. I replied that I had not many of them, had not yet made a sufficient study of the variety, and that I had all the time in the world for that.

Later on I had reason to regret this, because I lost the right of having raised the plant, and this is how it happened. In this spring of 1891 my violet 'Gloire de Bourg-la-Reine' was at the height of its popularity in the Midi. I had numerous orders from all sides, and I was unable to fulfil them. As I had a large quantity of this variety growing from seed, I sent them as 'Gloire de Bourg-la-Reine'. Now, it was among these same seedlings that the variety 'Princesse de Galles' appeared at Hyères, Nice and Saint-Raphaël[130], in a word along the whole coast. Many confused it with 'Gloire de Bourg-la-Reine', some growers persisted in calling it by this name. However, it is not the same; the petals develop more fully, the stem is slightly less rigid, and the violet-blue a somewhat darker shade.

'Princesse de Galles.'

In 1892 the few plants possessed by each of these growers and by me were propagated. At last on 23rd February 1893 I brought a collection of violets to a meeting of the Horticultural Society. Among them were three novelties, including 'Princesse de Galles' as No.1. I had decided to name two of these varieties at this meeting: one of them, No.1, to be dedicated to my daughter; No.2 – to the explorer Dybowski. But during the proceedings friends informed me that at the preceding meeting, on the 9th February, a florist from Hyères[131] (Var), M.Louis Achard[132], had presented a violet that was identical with mine. This was a disappointment, though not a great one. For the time being I gave up the idea of naming my seedlings.

There was only one thing for me to do; I had to know whether M.Achard's variety was really the same as mine, and whether it had become at all known. To this end I sent for some plants; it was exactly the same. What was to be done? I wrote to a few of my clients, who replied that it was making its way under the name of 'Princesse de Galles', that others called it 'Gloire-améliorée'[133]. Briefly, it would have caused confusion if I had renamed it myself. I had been at fault in not naming it at once; I just had to accept the established fact.

I have only given this detailed account in order to record beyond doubt that this variety first appeared at Bourg-la-Reine, long before it was thought of elsewhere.

However that may be, it is one of our beautiful varieties. It is, if not the biggest, at least one of the big violets ('Luxonne' grows as big, but the lack of firmness in the petals makes it inferior in the bunch).

With rounded petals, fully open flowers of a velvety violet-blue, it is a beautiful flower of fine effect. Its peduncles are quite strong and can be more than 20 centimetres in length, which makes it possible to arrange the flowers in all kinds of bouquets.

It is very free-flowering and does quite well when forced in winter; however, it must not be planted too close in the frames, for it then loses all its beauty and produces small flowers. This winter of 1895-1896 has not been good for 'Princesse de Galles'; all its flowers have been very pale.

The year 1893 brought many new violets, a most unusual thing with this race of plants.

I put on sale a variety with large flowers that I named

'Dybowski'. This variety, which is of a true violet colour with metallic reflections, is magnificent in bunches; the round, firm petals and the rigid stems make it very suitable. Its leaves and flowers are held up well, with peduncles of 18 to 25 centimetres. It is extremely easy to arrange in vases and make up into bunches, and it has a very pronounced and most sweet scent. The individual flower on its stem does not show up as well as does 'Luxonne' or 'Princesse de Galles', but in bunches it surpasses them.

The autumn completed the trio of these beautiful varieties with 'Amiral Avellan'[134], which was put on sale by Léonard Lille of Lyon[135].

This violet is quite remarkable owing to its novel colour, violet-purple with blue reflections in the flowers that are made up into bunches.(1) In the evening by artificial light it is a fiery colour, the bunch seems to be lit up. In brief, it makes a strange and beautiful effect, and has a sweet, most agreeable fragrance, which adds to the charm of this flower.

(1) I am referring to the large flattened bunches (montés plats), such as are delivered for soirées.

In appearance it is a plant in the style of 'Czar' as regards vigour and bearing. However, the leaves are a little bigger and darker and it has strong stems. The leaves and flowers are always held up well, the petals are fairly firm, of a round cup shape, and it looks well in bunches. It is a variety for spring rather than autumn.

I think I said previously that varieties that deviated from the colour violet were mercilessly rejected by the Paris market. I do not think it will be the same with this one. The few bunches that I have been able to make were all secured in advance and at a good price. I hope this state of affairs will continue, so as to defy routine a little.

Whether this be so or not, these three varieties are extremely interesting, as regards both flowers and leaves.(1)

(1) Speaking of leaves, I have just measured leaves of 'Gloire de Bourg-la-Reine' and 'Princesse de Galles' which were from 12 to 14 centimetres in diameter.

As I complete the biography of these few beautiful

varieties, the firm Molin of Lyon[136] announce two more large-flowered varieties. Since I have not yet been able to study them, I shall not give lengthy descriptions; and the few flowers that I have obtained this winter of 1895-1896 lead me to class them among the ordinary violets.

The first, 'Princesse Béatrice',[137] has average small rounded flowers: it is of the same type as 'Wellsiana', but flowers earlier, and is quite pretty. Its colour is dark violet, and it does not send out many runners.

The other, 'Comtesse Edmond du Tertre', also has flowers of average size, with very long petals of a much lighter colour. Briefly, it is a repetition of our fine 'Luxonne', and has the same defect, that of producing too many runners.

To sum up, they are two good varieties that have arrived too late.

As I am intent on drawing attention to all well-known varieties, I must also mention a white-flowered violet that has reached me from Italy under the name 'Princesse de Sumonte'[138]. Its white flower is lightly marked with lilac. Its shape is perfect, it is very fragrant, and well suited to forcing.

The note that accompanies this new variety speaks of it in the most glowing terms; may the future justify this eulogy!

The same is true of a very recent variety that I have just named and exhibited for the first time. This variety, 'La France', has the largest flowers of any violet up to the present time. The beautiful blooms always attain, and may indeed exceed, the diameter of a silver 5 franc piece.

Raised from 'Gloire de Bourg-la-Reine', it is a sister to 'Princesse de Galles', which it somewhat resembles in shape. It has large rounded petals, wide open throat; the spur of the calyx varies in appearance, being well pronounced and pointed in some flowers and almost non-existent in others. The peduncles are very firm, of a darkish green, and hold up their flowers well. Finally, it has an exceedingly sweet scent.

It differs above all from 'Princesse de Galles' in that the leaves are less erect. The plants being more compact and stocky, this allows the flowers to rise more clearly above the leaves. It is also more excessively prolific in flowering; it is not unusual to see hardy plants with eight or nine buds to each crown. Its colour, deep blue-violet with reflections, equally distinguishes it from its great sister. It makes a magnificent effect.

It is one of the most beautiful new flowers achieved so far.

Will it retain all these good qualities? I hope so. Here and now it has been admired by all amateurs at this year's concours général in Paris.

An account of Cucullata violets[139]

Before coming to the end of the description of violets, I must say a last word about four varieties which, although they have no scent, have made a mark for themselves. I want to say something about the four varieties of Cucullata, or perennial tuberous violets[140]. One kind, moreover, has been grown in France for a very long time.

It has been grown in the Paris area for more than thirty years, and the growers call it the "dog violet"; this is *Viola cucullata striata*. Although cultivators are apt to call it "dog violet", it has nothing in common with *Viola canina* of botanists.

Soon after it has started into growth it produces flowers; all the petals are striped with white. As it only blooms in spring, well after the last sweet violets, it does in any case find buyers. The foliage is abundant and very fine; the leaves are a delicate green, cordate and much serrated. Several score hectares of this variety are grown in the neighbourhood of Paris.

The second variety, *Cucullata grandiflora*, has a much bigger flower. Although it has been known for a very long time, the market-gardeners are not yet acquainted with it. If they had been, they would give up the first, the more so since this is much prettier. Its wide, well opened flowers are dark blue throughout, as big as those of 'Princesse de Galles', and make pretty bunches. It gives a prolific crop of beautiful large violets with peduncles of 20 to 25 centimetres. These two varieties are excellent for planting as a border.

The third variety of the genus is *Cucullata alba*, which has white flowers, as its name would suggest. This singular variety was put on the market by M.Dugourd of Fontainebleau[141].

Although very pretty, this variety has the same fault as 'White Czar' has among sweet violets; its flowers have little substance and are soon over. However, it makes a very fine effect, for it blooms with excessive freedom. Its foliage is also slightly smaller than that of its two fellow Cucullatas. But on the other hand it produces seed much more liberally. It is remarkable that since I have had this white-flowered variety, *Cucullata grandiflora*, which never set seed, now produces some.

These three varieties are stemless, they do not send out any runner or stem. They lose their leaves in winter. Each spring new leaves grow from the roots, which are woody rather than tuberous. In brief, this is a quite distinct race.

Some of the flowers have very pronounced spurs, others much less so. The capsules are oblong, they have three cells and three valves, which eject the seeds when they are ripe.

These three varieties appear to have come from North America, where they grow in the wild.

Finally, a fourth and last variety has just been introduced into Europe by the Forgeot firm: a violet with yellow flowers, *Viola pubescens*. This variety with stems (caulescent) comes from North America, and although its leaves have much the same shape and composition as the last three described above, it seems to form a variety of a separate kind. Its roots are not so woody, they do not form central rootstocks, and stems bearing flowers either settle on the ground or stand erect, according to whether they are long or short. As to the flower, it is a lovely dark canary-yellow, but very small, held on a short peduncle. It is not easy to pick, and may be considered useless for cut flowers. Like the others it is scentless. When well furnished with flowers it looks very pretty. It is a most novel plant for the amateur. Its capsules are, as with the others, oblong with three cells and three valves, which immediately eject the seed, in the manner of the pansy family.

Its general appearance does seem to set it apart from violets, and yet it can only be compared with the little *Viola biflora*[142], as they both have two-coloured flowers. Their leaves and flowers are practically identical, but *Viola pubescens* spreads out much more extensively[143]. Its flower has the perfect violet shape.

Here end the quotations, descriptions, presentation of good qualities and criticisms of the different varieties and species. From the various characters they possess they belong, as a family, to the genus Viola. But as I have not undertaken to write about the pansy family, I shall limit myself to those species and varieties which have nothing in common with the pansy, and which everyone at first sight takes to be violets.

Kinds of violets: violets with stems and without stems at the same time (1)

(1) I say with stems and without stems at the same time because in botanical description all sweet violets are classified as being without stems.

Well, that is not to be taken literally, for sweet violets do send out stems; they are all stoloniferous and put out runners or stolons which are at first only true, but not erect, stems; they produce flowers in exactly the same way as do true stems, until the stolon stops growing so as to form a proper crown. There are some which do not form crowns and remain very short. I have often noticed strong clumps made up of five or six crowns that sent out some fifteen stolons at least 20 centimetres long, furnished with buds and flowers. If you lifted up all these stolons, you would have veritable stems, independent of the crowns that remained at soil level.

The first kind is the true sweet violet, and all varieties that are derived from it.

The second, although many people confuse it with the first, is the unscented kind; and this also is made up of numerous varieties. I shall be told: "But in the wild it is confused and mixed with the sweet violet". I admit this, but what I have not had proved to me is that the two species hybridise. I do not know whether others have been more fortunate than I. In any case every time that I have sown sweet violets, though some had more fragrance than others, I always obtained sweet-scented violets. In the opposite case, when sowing scentless violets I have always grown violets without scent, never with scent. Moreover, when looking at the plants as a whole, the bearing and shape of these various scentless varieties seem to mark them out: the flower as a whole is flatter, the calyx more open, and the peduncle as it bears the flower is less curved. So in the woods I am seldom deceived and, without bending down, I could point out with my cane which are sweet violets and which are without scent, merely by glancing at the flowers.

All these observations taken together lead me to make a separate species of them, but I will not describe the method of cultivation, since they are not cultivated, or at least very little, and the method would be the same as for sweet violets. All that I can say is that in this species you find all the flawless tints of the violet colour, right through from pale blue to pure white, passing through twenty different shades, but that in unscented flowers I have never come across the intensely dark violet that exists with sweet violets.

The third kind consists of varieties of Parma violets. There too everything is different and distinctive; the leaves alone would form a species, if demarcations were established by the foliage. The leaves of all six distinct varieties known at present are almost alike: more incurved[144] than with odorata, the whole leaf is dentate and crenate, more elongated and pointed. The dark green is always glossy and much rougher to the touch. The leaf is moreover always held more erect.

As for the flower, it is of faultless shape, round for the most part; the petals, superimposed one upon another, form a perfect double flower. The colour passes from deep blue-violet, through violet, pale blue, reaches lilac-rose and ends at pure white. Finally, its scent is of the sweetest[145]. Whilst being similar to that of odorata, it is more delicate, more pleasing, more pervasive, without however being disturbing. Were it not for the scent, which is almost the same as that of odorata, one would wonder if this species really belongs to the genus viola or is a quite separate plant, a genus on its own. The seeds even, or to put it better the lack of seeds, would make one think so.

As I remarked in another passage, it cannot be said that Parma violets do not bear seed, since I have seen and demonstrated such seed, but that can be considered a phenomenon; and this lack of seed is what explains the few varieties that have been acquired up till now.

The fourth group, in my view, would be the woody kind, that is to say these three or four particularly remarkable varieties which, although perennial, lose their leaves entirely in winter, leaving, in order to survive this unfavourable season, only a rootstock consisting of an assemblage of little rhizomes as hard as wood.

The leaves develop in spring with the flower buds, and by the beginning of May they form beautiful clumps. But the flowers have no scent, and owe their place in commerce entirely to their having the appearance of violets[146]; in a word, they look exactly like them.

Their leaves are always heart-shaped and broadly dentate, of a fine dark green or delicate green. Rising to between 20 and 30 centimetres, according to the variety, all are uniformly without stem or stolon, and are increased by separating the little rhizomes. As regards colour, we have still only blue-violet and the fairly pure white. One of these varieties has white stripes down the blue petals and is free-flowering[147]. When all is said and done, these varieties from North America are very pretty and very hardy.

Finally, two autumns ago Forgeot imported from the same country a violet that is also of this kind, but differs in some points. The most unusual thing about it is that it has a yellow flower and that it is almost entirely composed of stems from which the small flowers grow on very short peduncles. The leaves are heart-shaped and dentate, as with the other plants of this group, but this plant has only one rhizome, or if there are several, they are not joined together, as is the case with those mentioned previously.

In brief, it is rather a novel plant than a very beautiful one. Like all violets, it actually flowers all the year, but from May on the flowers are not visible and immediately form a capsule containing seeds.

With regard to seed, all these varieties are each more curious the one than the other. Thus, if we admit this last one into the family, it produces seed in great quantity; the variety with white flowers also gives a quantity of capsules and good seed; only the freely striped variety never bears seed, although all four have single flowers.

PART TWO

**Cultivation of Violets
In Woodland, Gardens, Frames and Glass-houses**

CHAPTER I

Cultivation in bygone times

The common violet (*Viola odorata*) appears to be indigenous in our old Europe; it is found wherever the temperature does not exceed 30° or fall below -20° centigrade.

I have said on previous pages that all poets have sung its praises, they have all celebrated its charms, its intoxicating perfume, its modesty etc.

Botanists, for their part, have described the places where it grows. Some put it in shady and cool situations, others in places that are dry and warm. In my opinion they are all correct. It is found practically everywhere in cool, damp situations, in the shade. Then in another region one is amazed to find the same varieties where it is very dry and there is hardly ever any moisture. I have even picked violets in the mountains, in rock fissures which are rarely reached by any water.

All these citations would seem to imply that this plant has no need of cultivation. It grows successfully everywhere, produces a profusion of pretty little flowers that are the joy and delight of those who go to pick violets in the woods. I can well suppose, without insinuating anything unseemly, that in these little flower-gatherings one is not much concerned as to whether the plants have more flowers in one place than in another. However, this is always the case, and it does not escape the eye of an observant person.

And so, where violets are growing in shady places, they are well provided with leaves, and are fine plants of their kind, but they have the fewest flowers; whereas in clearings or at the base of sparse hedges you see good clumps all covered with pretty flowers. This belief that violets grow and bloom better in the shade than in bright light must then be considered a mistake.

It is strictly true to say that it likes semi-shade, especially during the summer months, as the sweet violet, although it is perennial, experiences a pause in its vegetation. Above all in

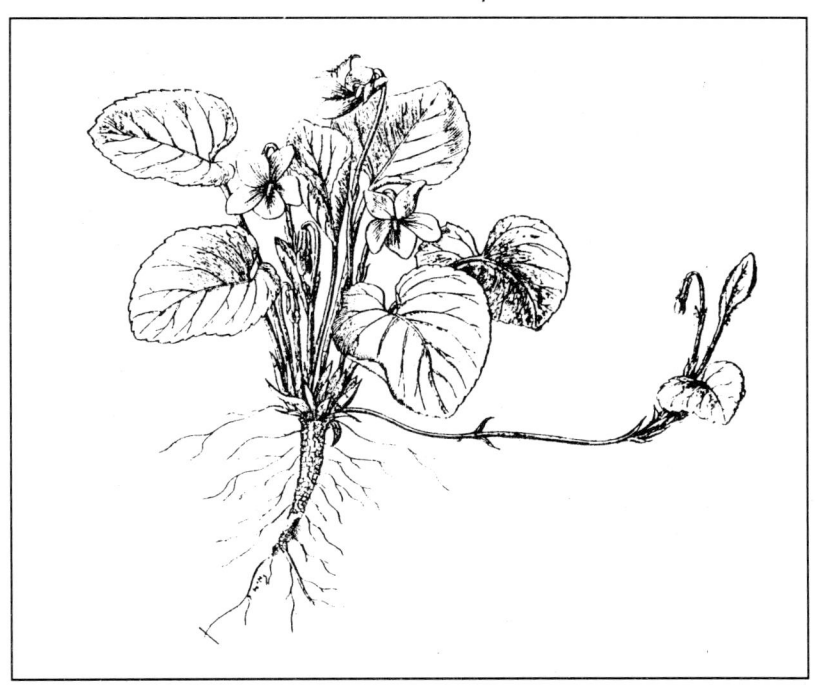

Viola odorata: Millet's illustration showing its habit of growth.

May and June, after flowering has ceased, it seems to stop growing. Then, in this latent state, it prefers to be sheltered from the sun's rays, which at this season are quite hot. Nevertheless, if this situation suits the plant as regards its foliage, it is not so for the production of flowers. The plant that has suffered a great deal during the summer will flower much more profusely in the autumn.

It is from observation of the modest wandering life of violets in the wild that I shall endeavour to give my readers some useful ideas for growing beautiful violets, under all kinds of cultivation and according to the requirements they have for these flowers.

From earliest times violets have been grown in gardens, but without any commercial motive. A few words that have come to us from history tell us that violets were cultivated in borders, but they do not tell us whether they were single or double, or what kind of violets they were.

Everything leads us to believe that, without more ado,

violets were taken from the woods, divided, then planted, and the job was done.

One must however take into account the fact that from the reign of Louis XIII there was great progress in gardening; but it must also be noted that this was above all among humble folk, in growing vegetables, the most useful kind of gardening for the family. The taste for flowers to be enjoyed for their beauty alone developed among the rich; but, given the lack of training of the gardeners, methods of cultivation and varieties of plants evolved slowly.

If there was a beautiful plant on one estate, a gardener or an owner would ask for it, grow it in his turn, and in this way it was established in one locality and did not find a place among the mass of people. One class of society alone had these privileges and enjoyed them; hence the slow rate at which beautiful things, new plants and the knowledge of how to grow them, became widely known.

Only our first revolution, having created equality of the classes, allowed everyone to see, I could say to relish, the floral refinements that lay dormant on the seignorial estates.

There was a new wind blowing, but only in the large towns. In the big centres people felt the need to enjoy the new life, and after so many disasters it seemed that one must steep oneself again in nature. Flowers are the first things to smile at us in these great cataclysms (1); and therefore many people went round the streets of Paris offering flowers to the citizens. Wild violets were among them, and gave a not inconsiderable reward to those who took the trouble to go and pick them and then sell them in Paris. It was merely a short step from that to growing the violets.

(1) Talking of this, the same thing happened in 1871 following the siege as took place from 1792 to 1800. As I grew flowers on a large scale, I consulted my father, who was my best adviser. In the spring of 1871 I said to him that it would perhaps be better to plant vegetables and other edible garden plants, which are more essential than flowers. But after we had talked it over I went on as before with my work on violets and roses. It was lucky for me that I did so, for these two flowers sold as though worth their weight in gold. During the autumn of 1871 and the spring of 1872 there was a sort of frenzy to have flowers. And yet the great city had only been deprived of these products for a single winter.

This is what happened under the Consulate[148]. Several small cultivators, who used to go to the woods to get violets, had the bright idea of taking plants in order to propagate them in small fields so as to have a more prolific and certain crop of flowers for them to pick than they gathered in one place and another.

Fresnes-les-Rungis and Romainville are two of the villages where violets were first grown commercially (1). Little by little the work became more remunerative, and this allowed for better cultivation. They began by making a selection of the violet plants; that is to say choosing flowers of a uniform violet colour.

(1) I say "commercially" deliberately. I repeat, for a long time already on large estates Parma violets had been grown in frames; and in borders the double blue and the single wild violet; but they were not put on sale.

According to accounts given by old growers, this is how they proceeded. The land was prepared as if for any vegetable, then four lines were drawn 30 centimetres apart. That formed a bed, and between each bed a space of 70 centimetres was left; this was the path. When fifteen to twenty beds had been prepared in this way, the violets were planted in them.

Little care was taken in planting out. As this was done in April, that is to say after flowering was over, they simply divided the old clumps and runners roughly and planted them with a dibber. The ground was then hoed a few times until autumn, and that was all. Moreover the beds were not renewed frequently; they were kept for four, five and even six years. During the first few years runners filled up the beds, which were merely weeded. With this method it was not possible to do more, so that in their last years in the bed the plants were poor, the flowers very small.

However that may be, with this unsound cultivation, the taste for and sale of violets increased, and along with this cultivation of native violets florists in the Paris area began to grow the Parma violet. It had been seen in various well-to-do houses, and it was a small step from that to people asking their gardeners to set about growing it. Produced on small plots and in small numbers, they were easily sold: Paris needed flowers. As for the way Parma violets were grown at this time, it was much the same as that used for ordinary violets. However, gardeners gave them shelter, brought them in during the winter and, in doing so, renewed them more often.

In the South of France Parma violets were grown solely for perfumery. The method of cultivation was practically the same.

As long as the First Empire lasted this cultivation made splendid progress; profits became more assured and people worked hard at growing violets: the plants were renewed at more frequent intervals, greater efforts were made to select superior forms of flower.

The sub-varieties became fixed and no longer resembled the original violets. The leaves were the same shape, the selected violets had a longer flowering season, and cultivation had extended: Fontenay-aux-Roses had added violets to its fields of rose bushes; Bourg-la-Reine and Sceaux were also growing them.

Unfortunately we come to the fall of the Empire; the disasters that followed put a stop to progress for a time.

However, from 1818 to 1820 in issues of *Le Bon Jardinier*[149] we read of violets being grown, double and single, and finally Parma violets, which, it is said, should be put in frames.

Going on a few years, we see a perceptible progress. It seems that on coming closer to Paris the fields of violets borrowed from the great city its refinement and desires. Plantations from seed, by chance, threw up sub-varieties of odorata that were more remontant, already giving some flowers in the autumn. At Fontenay growers, as a result of their close scrutiny, fixed late-flowering varieties which had flowers of a deeper violet, so that the period of harvesting was much extended and the crop much more remunerative. In brief, only the cold spells of winter stopped flowering. And so, to avoid this interruption, some growers started planting violets in frames.

From then on cultivation improved and reached a rational state which has not changed since, apart from the acquisition and improvement of new varieties. This will permit me to state the method of cultivation employed, and then the progress made as a result of experience gained during this time, and the need for the new varieties.

While setting out an account of the various violets, bringing them to the reader's notice, giving as well as I could their origin and place of birth, it was not at all easy to point out at the same time the different stages of their cultivation and the results they are expected to give, whether in amenity commerce or in industry.

I thought it would be much clearer and much to be preferred, when following this rapid historical account, to describe chapter by chapter the way in which the cultivation of each variety has developed and the situation that suits it: whether it be in woodland or large beds; in small gardens, on large estates or in parks; in glass-houses or frames; or whether, if we are dealing with large-scale commercial cultivation, it be in the open and with the use of hotbeds. I shall also show the different methods of growing Parma violets developed in the Paris area, in Toulouse[150] and along the Mediterranean coast.

CHAPTER II

Planting in woodland

In beginning with this kind of planting (I was going to say cultivation, but the term would be incorrect here) my aim is rather to teach what is to be done than to say how it should be done. As for plants, you have to act as in the case of gardens, a proceeding I shall speak of later on. Many people who possess woodland where there are no violets ask me to tell them how they can have plantings of this flower carried out. Nothing is simpler.

If the wood is comparatively big and contains walks two metres and more wide, you can establish borders consisting of two or three rows of violets, for nothing is prettier than that in autumn and spring.

It is sufficient to have the soil dug on each side of the avenue, 55 centimetres wide for three rows, and 40 centimetres only for two rows. Care must be taken to give the soil that has been dug a slight slope down towards the side of the avenue, and then a gentle blow to the surface of the soil with a fourche crochue (a fork with curved tines). The rows will be drawn out taking scrupulous care to follow all the contours of the avenue.

The distance between the rows should be 30 centimetres, if there are only two; 20 centimetres if there are three. As for the distance between plants, 30 centimetres for the three rows and 20 centimetres for the two rows. For narrow paths a single row is sufficient, but one should plant the violets closer together so as to form a compact line marking the sinuosities of the wood. Sparse plantings in woodland should always be made in clearings, in little groups joined together as I have said above. I shall not indicate here the method of planting, I shall simply point out the varieties that are amenable and will grow well there.

First of all the little Quatre-saisons violets, which are in their element there; then 'Argentiflora', a few specimens of which can be found in certain woods; and the little 'Viola odorata rubra' which, with its bright red colouring, does very well in a border and even in clearings. In a border you can very well have the three colours[151] by using 'Rawson's White'[152], which is only a slightly improved version of our white wild violet returned from England under this name. Two varieties that grow equally well under these conditions are 'Mignonette' and 'Lilas'.

Attempts have been made to grow large-flowered violets, including 'Czar' and 'White Czar', but without much success; and I have even seen Parma violets which, with the shelter given by trees, survived but only gave a few imperfect flowers. Generally speaking, double violets do not do well in woodland.

CHAPTER III

Growing violets in small gardens

Growing violets in small gardens is extremely easy and within the reach of all. The ordinary varieties are not expensive, whether you buy them or get them through friends, so it would be a great mistake to deprive oneself of this lovely little flower which, like the swallows, tells us of the return of spring, whilst at the same time wafting its sweet scent around our homes.

For these small gardens I would advise growing violets in a border, which economises in space and permits the luxury of having this plant without for this reason reducing the space kept for others.

These borders are always planted out in a row, the plants 15 to 20 centimetres apart; this distancing is necessary so that the foliage at once presents a becoming border. Planting is done with the dibber, using one-year plants, or divisions of two-year plants. The divisions must not be too big (two crowns or three (1), not more), and these crowns should be set precisely at soil level, and all at the same level.

(1) I say two crowns or three because I have often noticed that large clumps have been planted in order to have a fine show more quickly. This is a mistake; such large clumps waste away before new growth begins, the crowns weaken and finally are worthless and so do not flower.

The best times for making or remaking a border are from 1st October to 15th November or, in spring, from 15th February to the end of March. If you have to take plants from clumps that are in bloom, you must wait until the end of the flowering season: though the season is then less favourable for the resumption of growth, it will at least make it possible to pick the flowers!

If you want to grow your violets in a small bed it will be even better for them, but they must always be placed where there is fresh air and in the light. These conditions are necessary if they are to flower well.

The best varieties to use in small gardens (in saying best varieties I mean those that are the hardiest and have foliage that is firm and upright, making an effect) are:

CHAPTER IV

Varieties suitable for small gardens

'Czar', 'Quatre-saisons bleue', 'Wellsiana', 'Victoria', 'Luxonne', and 'l'Inépuisable'. These six varieties provide plenty of flowers in autumn and spring. Other varieties that also make very good borders, but which only bloom in spring are: 'Amiral Avellan', 'Viola odorata rubra', 'Bleue de Fontenay', 'Rawson's White' and 'Viola à fleurs tigrées or'[153].

Only two varieties of double violets are suitable: 'Double bleue' and 'Blanche double de Chevreuse'.

Although we have only a small garden, we may be a very keen grower of this flower and wish to have violets for some part of the winter.

To achieve this we will use the variety 'Quatre-saisons Semprez'. To begin with, about the middle of November, they will be planted in a bed in the garden. A suitable position, such as in front of a hedge or trellis facing south, is chosen to give protection; the plants are brought there with all the soil on their roots, and put in close together so that the leaves just touch. A few wooden laths are fixed above the bed, making it possible in severe winter weather to throw a mat or other suitable material across them to give shelter.

Barring a winter of quite exceptional severity, there will always be some violets in flower.

CHAPTER V

Violets in large gardens and on great estates. What can be expected of them and which varieties to use

The Head Gardener of a large estate is often hard put to it to provide sufficient flowers throughout the winter season: house plants, flowers for their scent etc. Well now, if the Head Gardener really knew how to grow violets and make them serve all the purposes that they are fitted for, he would certainly not fail to grow them.

I have seen many houses where the gardener delivered plants and flowers once a week. Among these flowers were often orchids, camellias, azaleas, and even lilac and roses, and then a modest little bunch of violets, of the ordinary single violets (quatre-saisons). You'll never guess what order was given to the gardener as he left: "Above all, when you next come don't forget the violets, if there are any". And this order expressed great longing; it made it quite clear that violets, despite their small size and modesty, were the favourite flowers.

Now, since as Head Gardener at a big house we have at our disposal the materials needed for cultivation, that is to say frames, mats and even dead leaves in the park for giving protection, we can provide violets every week from September right up to May. To do so all that is required is to have on hand varieties capable of serving this purpose.

In order to achieve this result it must be thoroughly understood that plantations of violets must be remade each year; that old plants must never be left when one is concerned with cut flowers; and only those in borders that are to flower in spring are to remain.

Let us then deal with the varieties that must be grown to get this result, which is a simple matter. Varieties planted in borders can provide us with flowers in autumn, that is to say September and October. For this we shall have the single violets: 'Wellsiana', 'Bleue de Fontenay', 'Czar', 'Lilas', 'l'Inépuisable'; and 'Patrie' as a double violet, if it is planted in fine sandy soil.

With the six varieties mentioned above we can have violets in bloom the whole of September and October. When winter is over these same varieties, still in borders, will of course come into flower again in March and April, as well as others that are specially intended for spring flowering.

Since we have flowers provided for September and October, let us prepare the varieties that are to succeed them; this is the most important season, all the winter, that is to say November, December and January.

For these three months a judicious choice is essential. Early-flowering varieties are required; when their growth is promoted and they are given protection, the most beautiful and fragrant of those that flower in the middle of winter will serve the purpose.

In order to get the result we are working for we grow them in open beds in the kitchen garden, just like strawberries, cabbages etc. "And why," I shall be asked, "grow them in open beds in the kitchen garden?" Quite simply because the violet is a plant like any other, and despite old prejudices which represent it as growing without the need of any attention, it requires care and preparation when it has a special part to play.

The best place for it is the kitchen garden, or the fleuriste[154] (an area reserved for raising flowering plants) if we possess one. Cared for just like any other plant or vegetable, it gets ready for a good period of flowering and forms fine compact and sturdy clumps, in a word clumps that are good for transplanting into the frames where we intend them to spend the winter.

Here are the varieties we shall use if we can, and if we have enough equipment to keep them all. As I give the approximate time and duration of their flowering, if the number of varieties is to be restricted, those that last the longest will be kept, or two varieties to flower in succession.

'Quatre-saisons odorante' violets flower from September until the end of January; 'Quatre-saisons bleue' from September until the end of January; 'Quatre-saisons Semprez' from October until the end of January; 'Souvenir de Millet père', which has large flowers, from November until the end of January.

Then there are the more unusual, large-flowered varieties: 'Gloire de Bourg-la-Reine', 'Explorateur Dybowski',

'Princesse de Galles', 'Luxonne', 'La France'. Flowering from November to mid-February, these varieties are of superior quality; they have very beautiful large flowers, long stems and lovely foliage. But they do not really bloom so freely in December and January as the four preceding varieties. In spring they can all perform well; there are only too many to choose from. Those that flowered in the autumn will start to bloom again. We can add to the number with 'Brune de Bourg-la-Reine' and 'Amiral Avellan', two superior varieties of purple and red, which, when made up into bunches and seen in a bright light, sparkle with a most effective fiery colour.

Then there are two hardy violets for the border in late spring: 'Rawson's White', a very appreciable pure white, and 'Odorata rubra', a purplish-red that is strikingly effective.[155] However, these two varieties have ordinary single flowers. Finally, if we have to make up borders that will be very attractive for the whole summer, above all in slightly shady situations, we shall have the violet 'Armandine Millet', which has green leaves variegated with ivory white, making a marvellous effect. It can also be accompanied by violets with leaves of a golden yellow, and also with very pretty stripes of the same colour. This variegation does not prevent these two varieties from producing in spring pretty blue-violet flowers above their foliage, which is in itself very striking.

Lastly, for lovers of double violets I will draw attention to 'Bleue double' and 'Belle de Châtenay', both of which are very late-flowering, flower profusely and have very full flowers.

To sum up, if one cannot, or has no wish to go in for all these varieties at one time, it will be enough to make a good choice from among them and by doing so one will obtain a fine succession of flowers from the end of August to the first days of May. This floral season may even be extended further by another kind containing three varieties, which will be dealt with in Chapter VII.[156]

CHAPTER VI

Parma Violets

So far there are but few varieties of Parma violets. In my opinion these few varieties constitute a species. Everything goes to prove it: its flower, its foliage, its scent, its constitution point to an origin other than that of central Europe. Is it really a violet? One might well doubt it.

However, the very few seeds that by great chance I have gathered are almost enough to prove my point.

Now what is its origin? This is far from clear: some say it came from Turkey, others say Parma; and then it also has the name Neapolitan violet[157]. Whatever the case may be, it can and should be considered a distinct species.

For those who love its sweet fragrance, it may be considered the supreme achievement of the whole race of violets. So if we are fond of its scent we shall want to have violets for seven months of the year.

Quick, let us raise Parma violets in one or two varieties to satisfy our requirements; they will not fail us. It is the prettiest plant of all. Just think, flowers during seven long months, and that in winter; what other plant would set itself up in competition for producing flowers over such a long period? Not many, I think. And so a good many garden owners give themselves this luxury of growing them; and but for a little snag this number would be even greater. This snag is that Parma violets cannot be grown out in the open. In winter they need frames, or at least the shelter of a wall to grow satisfactorily. The frame is above all its ideal situation. If raised in the open during the summer, then put in frames in October, it will be in its element there and flower abundantly for long months on end.

Where else can one find a plant that charms by its appearance and gives us scent as long as the flower lasts? And so we cannot be blamed for sheltering and pampering it a little in the winter; it well deserves such treatment.

In any case, whether we grow it with other violets in the

same place or on its own, we shall always be assured of a prolonged and fine yield of flowers. As I said at the beginning of this chapter, there are not many varieties of this species. At the present time there are only six: indeed, properly speaking there are only five.

'Parme ordinaire'. The typical or original form (pale blue-mauve).

'Parme de Toulouse'. An improved form of the preceding, a darker colour with firmer petals, obtained by cultivation and different climate.

'Parme sans filets'. A form of the ordinary Parma that does not produce runners.

'Mme. Millet'. Rose-coloured flowers; one of the most strongly scented.

'Marie-Louise'. Dark blue, with a small red fleck in the middle.

'Swanley White' or 'Comte de Brazza'. White, a very pure colour in winter; in the spring bright sunshine causes some tips of petals to turn blue.

As can be seen from this little list, 'Parme ordinaire' and 'Parme de Toulouse' are only forms of one variety. It is just a question of the selection of the finest plants grown in a climate better suited to its constitution that brought it to becoming a sub-variety. It would be perhaps ungenerous to reject it and confine ourselves to the type, the more so since it surpasses it in beauty, quality and yield. By leaving the list of varieties like this, confusion is avoided. The true 'Parme de Toulouse' is a much improved sub-variety developed over the years, and that is all.

That is why I shall refrain from registering names which are only repetitions of the Parma violet more or less well cared for, more or less well suited to the soils and climates where they are grown, such as 'Violette de Parme', 'Gloire d'Angoulême', 'de Turquie' etc. Some have not even been named.

Well, we have bought all these varieties and after growing them for two or three years we were convinced that this was really the true Parma violet with very slight variation according to its source and the climate of the place from which it came.

However that may be, and despite the few varieties we possess, as I have already said it is truly the violet above all others. And yet, before leaving the subject I must, for the sake of justice, address one more eulogy to it.

Everyone who has lived in Paris, even those who have merely passed through the city in March and April, have seen and admired those beautiful pots of violets which adorn our florists' shops. All the covered markets and street markets[158] are filled with them, flattering the eyes and filling the air of the metropolis with sweet scent. Well, it is the Parma violet again that supplies these quantities of plants. Not content with having paid its tribute in cut flowers during five to six months, at the end of spring it lends itself to this scheme, which consists, at the end of its flowering season, in putting three or four clumps together in a pot, then leaving them in a frame for seven or eight days, to give them time to recover and stand up well. They can then be enjoyed in the house until they have finished flowering completely.

This little operation, which is moreover very simple and will be dealt with in Chapter XVII, is profitable for growers, and is very useful for gardeners of private houses. As flowering plants are rather scarce in March and April, it enables them to supply floral decorations over a very long period. The rose-coloured Parma violet 'Mme.Millet' makes a pot plant with magnificent flowers.

CHAPTER VII

Late-flowering violets – Viola cucullata grandiflora and other so-called tuberous violets

This species flowers extremely late and is the natural complement to all other species and varieties. Its visible and beautiful flowers appear only in spring (I say visible and beautiful because, like almost all the violet family, it produces flowers practically all the year and very late). Parma violets alone still have a few blooms when it starts to flower. It is the ideal violet for borders: borders for large and small flowerbeds etc. It is hardy in all places, and grows as well in the sun as in semi-shade, in clumps that frequently carry fifty to a hundred blooms at a time. It is in bloom for a full month, and then its beautiful green foliage forms a pleasing decorative border for the rest of the year[159].

This species is not well known, is not widely distributed, not sufficiently so, considering the services it can render as a decorative flower. As a cut flower it makes very pretty bunches, after the manner of Parma violets; and moreover does well as a pot plant, especially if potted up when the buds are well-formed.

I said just now that it is not grown nearly enough. This is true, but it must be admitted that despite the size and beauty of its flowers, and even the length of the stems that make it easy to put these flowers in a vase, it has a major defect; it is without scent. For those who fear the smell of flowers in a room nothing could be better than to choose the varieties of this group, which can be used for all kinds of decorations without causing any inconvenience.

However, the new variety *Viola pubescens* cannot be included among those suitable for bunches, in which it would not look well. On the other hand, for putting in vases of various sorts or for flower arrangements in rooms, it produces a perfect embellishment. For a wide border, planted in the second or third row, the fine green foliage dotted with its many yellow flowers is tasteful and effective.

As opposed to other plants of its kind, it is not practicable to make increase by division. Stem cuttings do not give a satisfactory result. To make up for this it is quite easily reproduced from seed. From this year's sowings we have raised very fine plants.

Viola cucullata alba.

CHAPTER VIII

Methods of cultivation: preparation of plants

In previous chapters I have shown which varieties should be used, and now it only remains for me to make known the various operations that have to be carried out in order to bring to a successful issue a skilful cultivation of violets.

The sundry directions that I am about to give should apply just as much to the commercial grower as to anyone growing specially for a house. They can use the same methods, since they have the same end in view; creating fine plants able to produce as many flowers as possible. However, there are appreciable differences, sometimes with the one sometimes with the other, and this is the reason: the aim of the gardener growing violets for his own use is to have flowers at all times throughout the season; the ambition of the commercial grower is quite different. In many cases he will even sacrifice his plants; it matters little to him, since he will propagate others, provided that he has his flowers at the expedient time, that is to say in time for the holidays, when they fetch a considerably higher price. Achieving this result depends more on the way the plants are forced than on the way they are raised. Let us then turn our attention to the raising of these plants.

As I have already observed, violets, especially the singles, adapt themselves to almost all soils, but prefer land that has been prepared well in advance by winter digging and suitably, but not excessively, manured. Violets do not like too much manure; they do not succeed at all on the land of market-gardeners, where manure forms the greater part.

The soil having been worked thoroughly in this manner, we divide it into beds, which are prepared as if for any other vegetable or plant; that is to say the surface broken down with a fourche crochue and raked. Finally four rows are drawn out along it. Our rows will be 40 centimetres apart, which will make our bed 1m.20 wide. Then our plants will be set at 30 centimetre intervals.

These instructions apply only to varieties with large foliage, such as 'Gloire de Bourg-la-Reine', 'Princesse de Galles', 'Explorateur Dybowski', 'Amiral Avellan', 'Czar', 'Brune de Bourg-la-Reine', 'Wilson' etc. For the ordinary varieties with small leaves such as 'Quatre-saisons ordinaire', 'Viola odorata rubra' etc. the rows will be spaced no more than 30 centimetres apart, which will make our bed only 90 centimetres wide. We will also put our plants a little closer together; instead of 30 centimetres we will set them 25 centimetres one from another. These distances will suit our violets in both cases.

If we wish to make several beds of violets, and this will be the case with those who grow them on a large scale, we will make paths of 60 centimetres between each bed. At first sight this may seem wide, but when the leaves of the violets extend 10 centimetres from each side, our path will be reduced to 40 centimetres, which is certainly not too much for what will be required of it during the season's work, and even the whole year if the varieties that have been planted are to remain in the bed and flower there.

I have just stated the distances to be followed when planting, without even indicating what plants are to be used. Well, dear readers, the plants we shall use will be small clumps consisting of one crown or two crowns at the most, with young roots, plants which will have been prepared for this purpose during the preceding winter. If we do not have these fine little plants, let us use divisions of old clumps, consisting also of two or three crowns, as well rooted as possible. But in any case these will not be as good as plants specially prepared for the purpose. The preparation of these plants will be considered at the end of this chapter.

As we shall be using the dibber, planting must be thorough and careful. Generally speaking plants from old clumps will have long roots, and a fairly deep hole will have to be made with the dibber so that the plants go right down to their full depth and can then be firmly settled at once; otherwise the plants are in jeopardy and our plantation endangered. For, contrary to the usual view, violets, especially in spring, are difficult to get into growth again and there are few plants which, set out at this season, remain for so long without putting out new roots. Planting out will take place in March and April, but not later than the latter month, especially for single varieties.

Since our violets do not change, many making hardly any progress for a long time, we will avoid disturbing them, that is to say hoeing them straight away. If weeds appear it will be preferable to hoe them up, but in June hoe shallowly, making quite sure not to touch the newly grown roots. At the end of July we will give a second, somewhat deeper, more thorough hoeing; and yet if the weather is very dry it will be better to refrain and delay the operation; our violets having begun to grow more vigorously may make this possible. I also forgot to say that June and July can be very dry; despite this do not water your violets before 15th July. Before this is not the season for strong growth, and one would do them much more harm than good. It is permissible to start watering them at the end of July, but this must only be done if one is determined to continue should the autumn drought persist; otherwise they would be aroused into active growth, and passing from one state to another could be seriously harmed.

During August the violets spring forward and start growing vigorously. Towards the end of this month we destroy any weeds there may be and remove the runners or stolons, which will already have begun to grow. This slight work is necessary if we wish to have beautiful flowers. I am well aware that you could say: "But I have never taken runners off my violets, and I have had flowers all the same". I know that very well, and the same goes for everything, flowers, fruit etc. But since in this matter we wish to have something beautiful, let us work to achieve it. Moreover these little attentions do not take long, and so towards 20th October let us remove the runners for a second time. This serves two useful purposes: the first is to relieve our clump of all that could take away sap to its detriment; the second is that, while growing vigorously, our clump has sent out fine stolons that we are going to make use of. Now is the moment to think of providing ourselves with plants for the next year. It is not difficult; instead of disregarding the runners that have to be removed, as we did the first time, we collect them as the work proceeds and put them under cover so that they do not wilt.

When all runners have been removed, we proceed to their preparation. This is quite simple. They are arranged with the terminal crowns all facing in the same direction and the little sidegrowths are carefully removed. They are then taken in handfuls and made of equal length with a pruning or grafting knife, so that they are uniform for planting. It now remains for us to plant them or heel them in in a nursery bed to pass the winter. For this purpose we will prepare a bed of good, fairly light soil, and preferably with a favourable aspect. We will put together three or four runners and plant them close together, that is to say about two hundred little clusters of three to five crowns per square metre. Others are content to heel in these plants, laying them almost side by side, with the small propagation rows spaced 15 centimetres apart.

In any case I prefer planting them, it is more certain, but it does perhaps take a little more time. Finally, to get through the winter the plants must be given a light covering of some kind or other, only during severe frosts, so as to save the leaves from scorching.

With the return of spring our plants will start into growth and make new roots. We shall then simply have to take them when planting our violets in March and April. As was said before, many people do not give themselves this trouble and take divisions of old clumps. The latter method is quite successful, but is never as good as having fine plants prepared in advance.

CHAPTER IX

Cultivation of Parma violets and hardy double violets. Preparation of these divers plants.

Preparation and cultivation of the various varieties of Parma violets and almost all violets with double flowers differ little from the methods employed with single-flowered violets. However, they require some special attention.

Ground is normally chosen where the soil is lighter and more friable than usual. Without adding an excessive amount of manure, the soil should be improved with some good humus, either horse manure or leafmould; one or the other will be very beneficial for heavy soils. If we are bent on growing Parma violets, we must add to the soil sand, old rotted turfs, in a word all the composted material that can make our heavy soil lighter and more porous.

Planting will be done in the same way as with single violets: always with a dibber, but the small volume and fineness of the leaves will enable us to plant more compactly. Instead of drawing out four rows in our beds, we will draw six rows, and instead of putting our plants 30 or 40 centimetres apart, we will put them 15 centimetres apart at the most. This may seem to give too little space for the plants, but it is not so at all, for it only means forty to fifty clumps per square metre, quite enough space for Parma violets. The same distances can be used with the double violets 'Patrie' and 'en arbre'[160]; but for other double-flowered varieties I suggest having only five rows and setting the plants 20 centimetres apart.

Since we are in the chapter dealing with tender violets, I think it would be a good idea to bring under the same regime two very pretty violets, the varieties 'Armandine Millet' and 'Souvenir de Millet père', which, when grown like Parma violets, give very fine results.

When we have finished planting our Parma violets, following these directions, we shall have nothing more to do other than to give the same attention as has been set out for single violets, that is to say weeding, hoeing and removing runners. However, if our soil is dry and scorching, a little mulching with very fine material, or better still coarse compost scattered broadcast with a shovel at the beginning of June, will be of great benefit to help us get through July and August. In this latter month our violets, principally Parma violets, will have reached the stage for derunnering, which must not be neglected with Parma violets, as they have a tendency to produce a great many stolons. If these runners are not taken off, the crowns waste away and produce very weak flowers. Moreover, as these violets only give good results if they are renewed annually with young plants, let us make use of these stolons to prepare plants for the next year. This will be done in the same way as for other violets.

Having arranged the stolons in handfuls, that is to say equalised them, you place all the crowns at the same height so that they can easily be taken and planted. Whilst carrying out this slight adjustment you remove the sidegrowths at the base of the runners so as to expedite the appearance of rootlets. Once our plants have been prepared in this way, they are set out in healthy light soil in a good situation. In advance a bed of soil the width of the frames will have been prepared; for, in order that they survive the winter, it is essential to put Parma violets in frames.

They can be just poor frames that have been discarded; provided that they give protection during the worst of the winter that is all that is required. I insist on the use of old frames because it is a way of using them and keeping the good ones for other cultivation. Our specialist growers use no others, and as they need thousands of plants, all their old material goes to provide this shelter.

The plants are very close together in the frames. No fewer than 250 to 300 little clumps are planted in each frame, and each little clump is itself composed of four or five runners or stolons. Once this planting has been completed hardly any further attention is called for until April and May, when one comes to fetch them for planting out in the open. Briefly, they only require to be kept free from weeds, as is the case with any plant, and this done, we shall have fine plants at our disposal in the spring.

However, if when runners are removed in August there are insufficient for our needs, we can make up our numbers at the second derunnering. Those taken at this second operation are a little weaker, but they are excellent all the same.

I say second derunnering: this is of course a trifle burdensome, but necessary; for we all know that with the climate of Central France and even almost everywhere, Parma violets must be brought in for the winter. It is a few days before they are put under cover that the runners are again removed.

These then are about the only little tasks that Parma violets and a few hardy double violets necessitate, from when they are planted in the open until they are put under cover; that is to say the moment when, having been put in their place, they begin to flower.

It is also necessary to say that, as with single violets, if the months of August and September are dry, they must be watered well and continuously.

Before leaving this chapter it will be convenient for me to mention the method of propagating the variety 'Parme sans filets'. Since it does not produce runners, there are none to be removed in the season of vigorous growth. Many growers restrict themselves to dividing up the clumps in spring in order to increase their plants. This is good, but another method is even better.

I said just now that plants of this variety do not have stolons. This is true; but they form clumps by the joining together of numerous crowns attached on the roots at a slightly greater depth than in the other varieties. I advise then that when the violets are brought in for protection, three or four of these little crowns should be detached, especially from strong clumps, and replanted in a frame, just as with the stolons or runners of other varieties. I recommend above all that when they are planted the neck of the plant should be well firmed in, in view of the absence of a long stem, such as the stolons present. When they are prepared like this we shall find in spring good handsome little plants of 'Parme sans filets' that have not flowered, ready for transplanting. These little plants or crowns will be far preferable to divisions made in spring, which will already have flowered.

The violets 'Patrie', 'Blanche de Chevreuse' and 'en arbre' produce few if any runners, and so one can use the same method for these varieties as for the preceding one.

CHAPTER X

Propagating and cultivating *Viola cucullata*

To the best of my knowledge there are not many varieties of this unscented species, in France at any rate. I have only three: *Viola cucullata grandiflora, Viola cucullata striata* and *Viola cucullata alba.* I should say that as regards roots or basal growths they are practically identical. Although the leaves of all varieties are cordiform, there is considerable difference between them. *Cucullata striata* is the first of these varieties to flower. It makes strong clumps, the delicate green leaves reaching a height of 30 or 40 centimetres. Then comes *cucullata alba,* whose foliage is smaller; a strong green, but almost white where it joins the stem[161]. *Cucullata grandiflora* is the last to appear; it has dark green leaves on shorter stems. Its flowers rise above the leaves, whereas at the end of the flowering season the other two varieties have their flowers hidden in the foliage.

As an ornamental plant in gardens *cucullata grandiflora* is truly worthy of interest. When picked the flowers look very pretty in vases in the house, since they are large and the stems long.

Propagation of these three varieties is carried out either in spring, before they flower, or afterwards in autumn. I prefer to do it in spring.

This species, as I have already said, is fibrous, woody. A strong clump consists of several of these roots resembling little pieces of thick stem with a gnarled trunk, assembled next to each other and so forming a kind of bulb, whence the name given to it by some people, "tuberous violet".

Their propagation is very easy, for these plants do not put out runners or well marked stolons; one has only to divide these clumps into three or four smaller ones with the end of some sort of spatula, so as not to scrape the rhizome or pull off roots[162]. Once they have been divided, it only remains to plant them with a trowel as one would any other kind of plant.

As these varieties are late-flowering and we have no intention of lifting them to put them in frames, we will try to plant them in their permanent position, whether in a border, in beds, or even as a decorative feature in a massed flower bed; no matter where, provided that they are not disturbed when they come into flower.

Although truly perennial, these violets do not keep their leaves all the year; they grow from the spring, form very beautiful borders, and then disappear in the autumn. So we must take precautions to provide against this lack of greenery during the winter. As for the attention they require, this is even easier than with the scented violets. Once they have been divided and planted out they only need to be kept free from weeds throughout the year. So far the only variety we have successfully propagated from seed is *cucullata alba*. (For propagation of the yellow variety see page 95.)

CHAPTER XI

Single sweet violets in the winter season; method of forcing; the attention they require

In a previous chapter we were concerned with propagation and how to get the violets into vigorous growth. We have now reached autumn with fine clumps of this year, ready to yield a good harvest of flowers, because they are young plants.

For those that we have planted in a border nothing further needs to be done, except to enjoy them so far as the vagaries of the weather allow. For those in beds in the kitchen garden, if we have no frames or sheltered places we shall be compelled to do the same as for the borders. However, as good strains bloom from September and continue flowering in October and November, we ought to try and make a little sacrifice so that the time when no flowers are produced should be as brief as possible; and as what is not possible for the borders is possible for the beds where they are gathered together, during the severest winter frosts we will throw a few armfuls of straw or hay lightly on them, so as to prevent hoarfrost or ground-frost from cutting back the foliage. If there is a covering of snow on the ground there is no point in covering the plants, as in this case the snow itself is sufficient protection. As soon as winter is past we take off the covering, our violets start into growth again, and continue flowering all the spring.

In other cases, if we have the material, that is to say a few frames, at our disposal and we are required to supply violets throughout the winter, we shall be compelled to put the plants in frames. This is easily done, for, as we know that our violets have been properly cared for, their runners removed at the right time, we have only to take our fine clumps and plant them in frames.

Before doing this there are a few little preparations we must make. As it is for the winter period, we are careful to turn our frames so that the base faces south, having a decided slope, in order that the rare rays of sunlight should strike our panes of glass. When this has been done, the soil in the frames is thoroughly lightened and made finer by the addition of compost or sand, depending on the condition of the earth when we are working it. Then it is carefully divided within the frames and made quite level; I say level, but it should have just about the same slope as the frames, so that the plants can enjoy what sun's rays there may be. Once these little preparations are completed planting takes place. However, if we have to provide violets during the whole season, we must choose our varieties with discernment.

The varieties 'Quatre-saisons Semprez', 'Quatre-saisons ordinaire', 'Lilas', 'Souvenir de Millet père' will be the mainstay for the depths of winter; 'Gloire de Bourg-la-Reine', 'Princesse de Galles', 'Dybowski' will follow; and finally 'Amiral Avellan' and 'Czar' come after them, leading on to those grown in the open.

Having made our choice, we plant 45, 56 or 64 clumps in each frame (1) according to their vigour (you can even put more in if you so wish). To begin with the violets are taken up with a good ball of soil attached; the old leaves, as well as those that have turned yellow[163], are removed; then, using a small trowel, we plant the violets in their permanent positions.

(1) I refer here to the normal forcing frame; those most in use at Paris are 1m.40 long by 1m.30 wide. All proportions must be kept if the frames are a different size.

When all this is done the frame lights are put on, and a good mat on top if the weather is cold, and then we are assured of a fine crop of flowers during the winter. "But what", you will ask, "if the cold is intense? Since we have arranged our planting in open ground simply in frames with top lights, won't our plants get frosted?"

That is true, but as I have spoken of giving heat I shall explain myself.

Actually violets do not require heat, and stove-house heating does not suit them. Later on I shall say a few words on growing violets in a greenhouse, but for the moment let us return to the practical work, that which will enable us to have an abundance of beautiful flowers at little cost; and this is not difficult with violets.

As I said, violets do not like a lot of heat, and it is enough to have the space for making a slight source of heat in the ground on each side of the frame. The soil of the paths around our frames must be dug out two spits deep and the trenches filled with leaves or horse manure well compacted. This will produce a slight heating throughout the soil and the plants, which will be sufficient to keep the violets in full activity. If the winter persists, in extremely cold weather we shall renew this heating by bringing some more manure or fresh leaves, which will be mixed in again and well trampled down, and this will continue to maintain the slight gentle heat that the violets need.

As for violets of the second and third season, it will be sufficient to put a mat on the frame during very severe winters. Even when covered in this way they may get frozen, but this will not harm them when the frost and snow melt.

Immediately mild weather returns they start into growth again and commence flowering throughout the spring.

Treated and heated in this manner they are bound to give a satisfactory result.

Apart from putting on and removing the covers each day, there are few daily tasks during this long period of time. There is, in fact, very little to be done, and here it is in a few lines.

Once the plants are properly settled in they should be watered copiously; at least two waterings, which will serve to wash down the foliage as well as moisten the soil. In mild weather the frames must always be fully ventilated so that the violets do not wilt and the flower buds fail to develop, as sometimes happens. When too much heat is given, the leaves grow a delicate green, become very drawn, and the few buds that are fully developed come into flower; but those that follow are aborted and remain in the crowns, putting an end to the premature crop. With a little experience this can easily be avoided, and, knowing that violets do not want a great deal of heat we will not force their growth in this way more than the season compels us to.

To sum up, what is needed is a lot of light (the plants should be very close to the glass); good ventilation; and a quite moderate temperature. There is also one little job to be done at least twice during the flowering season. We all know that although violets are perennial and in leaf throughout the year, they lose leaves from time to time. And so, a month after they have been put in frames a good few leaves turn yellow as a result of being transplanted. These should be removed, as this gives air between the plants and maintains healthy vegetation. In many years this operation must be repeated twice.

CHAPTER XII

Forcing Parma violets; the results that can be obtained from this

All that has been said about putting single sweet violets in frames can be applied to Parma violets, with a few slight differences. We know that Parma violets are a little more delicate, and so we shall act accordingly. The soil in which they are to be planted will be very light, and sandier if possible, and the planting (1) should take place a good month earlier than with ordinary violets.

(1) Planting Parma violets should almost always be done with a ball of soil on the roots of the plants. During the last few years I have gained experience of the comparative results: those planted with a ball of soil did not suffer any check, and as a result did not get mildew in the winter; those that were planted with bare roots suffered a severe check and in the following months were greatly affected by this. This has been the case each time I have made comparative trials.

So planting can commence from 1st October. Parmas are the preeminent violets for frames, and are not fond of frost. One can also put more plants in a frame, and instead of sixty at the most, eighty to a hundred are possible, according to the strength of the plants, with the exception that there are sandy and peaty soils in which Parma violets race away and become very vigorous. Then not more than sixty to seventy must be put in if one is not to run the risk of having a muddle in the frames. They should always be very close to the glass, from 10 to 15 centimetres, not more.

Having watered the plants well immediately after planting them, we can relax for a good month.

All Parma violets are very suitable for forcing, in the climate of Central France, and it is even necessary if one is to get the maximum yield.

'Marie-Louise' and 'Comte de Brazza' would be the two

most difficult varieties; especially 'Marie-Louise', whose foliage is a little more delicate than that of other varieties, and is more subject to mildew, particularly in Paris, especially if it is not planted with great care and if it suffers a check when transplanted. As for 'Comte de Brazza', it grows and reacts well to forcing, but being by nature late to come into bloom, it produces fewer flowers in November and December.

We use the same method of providing heat as described before and dig out trenches around the frames, only a little bigger, since Parma violets are fond of warmth, and they will like this. In the case of an extremely mild autumn we will give full ventilation, and even remove the lights when necessary, temporarily of course. On the other hand if the weather is harsh we will renew our paths each month so as to maintain the heat.

Treated in this way Parma violets can and indeed should produce flowers for seven consecutive months. They have one other undeniable advantage. Many people do not live in the country in winter and yet wish to enjoy their Parma violets. Now, besides picking the flowers, which can be done every week, you can pot up this charming plant, some ten days before taking it away for use. With a few pots you can adorn your dwelling and fill it with scent, a genuine advantage in winter.

I should cite one little treatment of Parma violets which, though hardly ever employed now, was carried out for quite a long time by gardeners of private houses and horticulturists and which, it may be said in passing, gave very good results: the plants were stronger, the flowers larger and with longer stems, only there were fewer of them per frame. This is what was called forcing on the spot.

This is what was done. In spring beds of soil the same size as the frames were prepared; then seven or eight rows were planted so as to have fifty-six to sixty-four plants per frame. All summer they were given the same attention as indicated previously, and in the autumn, after the last removal of runners had been completed, instead of taking up the plants, the frames with their lights were brought and placed over the violets, so that the plants received no check whatever.

After this, heating was provided by the paths in exactly the same way as when plants were moved into the frames, and a good crop of beautiful flowers was obtained. The reason for abandoning this method, at least on the part of horticulturists, was the lack of success in raising plants. In fact it often happened that there were many plants missing from each frame, they had to be replaced, and this was not very practicable; or else the clumps were not strong enough and did not fill the frames, which had to be heated all the same. This was not the sole reason, however, for the abandonment of this method of cultivation. Year by year the sale of Parma violets in pots became more marked. Those that had been transplanted in autumn were better adapted to being put in pots and did not suffer from this, whilst those that had not been disturbed and had all their old roots were not suitable at all. However, in private houses, where Parma violets do well and where they are grown exclusively for their flowers, I would still advise this method, which is very simple and inexpensive.

CHAPTER XIII

The forcing of Parma violets and single violets, as practised by specialists in the Paris region

Cultivation and propagation of Parma violets, as well as that of single and hardy double violets[164], is practically the same everywhere. In Central and Northern France, and abroad, it is partially or wholly neglected, and the few firms that attempted to set themselves up in this trade, not having either the principles or the materials needed to carry it on well, gave up after a few fruitless attempts. Until now foreign countries are dependent on the Paris markets and those of the Midi. England, Belgium, Germany, Austria and even Russia buy these flowers from France, either in Paris or in Provence.

In the Paris area there are specialists who engage in this work, so specialised that those who grow Parma violets do not grow single violets (1) and vice versa. At the present time in Paris more than twenty thousand frames are still taken up with the forcing of Parma violets, and that despite the consignments from the Midi, which considerably decrease the profits of the Parisian growers. Were it not for the disposal of all their violets sold in pots they would not be able to stand the competition.

In a later chapter[165] I shall speak of the cultivation of violets in pots.

(1) With regard to single violets readers may wonder why I always speak of single violets and Parma violets, and when talking about cultivation I never mention hardy double violets. There is a logical reason, which is that they are not grown commercially: double violets other than Parmas are not appreciated as popular cut flowers. Many times attempts have been made on a small scale, but without success.

However, a village near Paris called Chevreuse has for many years supplied Paris with double violets, double white and double blue, not for cut flowers but for planting in small gardens. They were conveyed as little clumps with a ball of soil, in wicker baskets, each of which held twenty-five or thirty.

The forcing of violets by specialists

Today this work has practically ceased, not that there is no demand in spring but, according to what the growers of the region say, the violets no longer grow well enough. Briefly, that is all that was grown by way of hardy double violets in the Paris climate, and that simply in the open.

As I do not wish to retrace my steps and since new plants of violets are raised in much the same way everywhere, especially by specialist growers, I shall not deal with this matter here. But I must say that, in view of the vast numbers that are grown, they are all raised outside the gardens[166], in areas where they are attended to daily, above all for removing runners. There are two main reasons for these derunnerings: the first is that plants are needed for the following year; the second is that a plant whose runners have been carefully taken off is much more suitable for planting in pots.

As soon as all the violets have been planted in their permanent positions – and this should be completed by about 1st November at the latest – the good cultivator at once begins to apply heat from all the paths between his frames. These paths have been excavated 45 to 50 centimetres wide and to the same depth. They are filled with horse manure in such a way that after this work of filling the paths is completed a block of 500 to 600 frames, or even more, presents a level surface with little for the frost to seize hold of. Afterwards it is sufficient to cover the frames with one or two mats, depending on how low the temperature falls and the severity of the winter. Every three weeks the paths are renewed; a little fresh, warm manure is added, and this reactivates and continues the slight heat. This process goes on until the end of winter, the time when the violets are to be potted up and sold.

I have spoken of the attention that must be given to maintain the health of the plants when growing on a small scale. The same work is necessary when growing on a large scale; however, less ventilation is given, since it is desirable to keep as much heat as possible when bringing the plants on more rapidly.

The number of specialist growers of single violets is tending to decrease rather than increase, especially in Central France and in the Paris area. The effect of flowers from the Midi is felt even more on this kind of cultivation, presenting formidable competition in sales. The former high prices are no longer obtained; 4 and 5 francs are the highest rates received for the large flat bunches of single violets, whereas about ten years ago as much as 7 and 8 francs were obtained. This fall in prices, reducing profits, makes competition almost impossible; and so from 25,000 to 30,000 frames devoted to this cultivation only 15,000 to 18,000 are in use for this purpose at the present.

The horticulturists and growers engaged in this branch of the trade treat single violets in exactly the same way as Parma violets. They are almost all raised outside the gardens, then brought and planted in frames in the autumn, a little later than Parma violets, and the heating is a little less serious, as they flower more vigorously in mid-winter than Parma violets, and this allows the grower to respond to changes in temperature. So in a mild winter only a little heat is given; if on the other hand the winter is severe the paths are stirred up so as to give more heat and avoid any interruption in the yield of flowers.

The varieties used for these various forms of forcing are few in number. Until 1870 early-flowering quatre-saisons violets were employed. Then two varieties were cultivated: the violet 'Quatre-saisons Semprez'; and from about 1873 many adopted 'Czar', popularly known as 'The Russian'. Since this period the varieties used for this purpose have remained the same, and they still make the major contribution. However, I should note that for the last three years some horticulturists have added the large-flowered 'Luxonne' to their work. The variety 'Princesse de Galles' too has been used by some firms, and I hope that the same will happen with 'Amiral Avellan' and several other very beautiful violets.

I could not say with certainty that these magnificent varieties are going to give a great boost to this peculiarly Parisian method of cultivation; however, they will make it possible to compete on equal terms in supplying bunches of violets of superior quality, enabling the growers to charge remunerative prices.

CHAPTER XIV

Violets grown in the open; their cultivation and development

As with frame cultivation, so the art of growing violets in the open has been thoroughly attained by their rivals in the Midi, and the growers of Paris have been compelled to seek a way of lessening their trouble by changing varieties.

Cultivators of no more than ten years ago grew 'Quatre-saisons ordinaire' and 'Quatre-saisons Semprez' for the winter season. Although they did not give the violets any protection, they nevertheless tried to get the maximum yield. If the winter was mild, they benefited from this and had a good season. If on the other hand it was harsh and severe, higher prices for their few bunches got them out of their difficulty. Today this is no longer the case, and these same bunches of imperfect violets (because they were not protected) are no longer bought, and the rival from the Midi shows at its best.

Since then the growers have sought another solution. They noticed that in the autumn, when the weather is still mild, and also in warm spring weather, violets from the Midi arrive all curled and in a poor state. In order to profit from their rivals' disadvantage they tried to find a variety which, whilst still remontant, had mainly two good seasons of flowering: one in the autumn and the other in spring. This they achieved by means of selection, with the result that there are hardly any flowers to pick in winter, and on the other hand those of the autumn and spring are of superior quality. This has entailed the fatal loss of the cultivation of 'Quatre-saisons hâtive' and 'Quatre-saisons Semprez'; a few hectares here and there of 'The Russian', and that is all that remains of the varieties of these past years.

If we were to go back thirty years we would find the greater part of these violets being grown at Fontenay-aux-Roses, at Sceaux, and especially at Verrières, and finally at Palaiseau. Today they are still grown in all these communes, but far less. The centre of this activity has moved, and you have to go beyond Palaiseau to find violets grown on a very large scale. Marcoussis and the surrounding area are covered with these plants, more than a hundred hectares of them are grown. Towards the end of the Empire, about 1867, there were at the lowest estimate two hundred hectares.

These violets are cultivated in the simplest manner; no derunnering is carried out, and the plants are propagated by dividing the clumps that have flowered.

No good cultivator leaves his violets to flower more than one season, from September to April. When by chance the clumps do not become strong enough however, he leaves them for a second season. To this end he carries out the following little operation. Some time after picking has finished, about mid-May, with the use of a pruning knife he cuts off the runners, decreases the size of the clumps, reducing them to five or six crowns. After this they are treated as if newly planted.

For these plantations of violets large pieces of land are deeply dug, a good spit (1) deep, then manured moderately, on the surface or dug in as the digging proceeds. This work is followed by breaking up the soil by hand with the fourche crochue (2).

(1) I say "fer be bêche" (spit) to show the approximate depth; growers in flat open land never use the "bêche" (spade), they do this breaking up of the soil with a "houe" (hoe), a tool similar to a "crochet plat", only very curved and with a very short handle.
(2) Our growers in the Paris area use a large iron six-toothed rake, which is popularly known as a "galère"[167]. This tool does a finer job than the fourche crochue and answers two purposes: that of breaking up the soil and that of raking. It is very practical for cultivation of the fields.

The land is then divided into beds one metre wide, which are formed into four rows. The paths are 60 centimetres in width, and this is reduced to 40 centimetres when the clumps and leaves are large; this is not excessive when cultivation is being carried out.

When the beds have been marked out in this way, the next stage is to plant the violets. Clumps of the previous year are taken up, divided into pieces of one or two crowns and then planted with the dibber 30 centimetres apart. The hardest part of the work is now completed; a weeding one month later, then hoeing twice during the course of the summer, and we come to September, when we start to pick the flowers.

In connection with the "harvesting" of violets I should say that, just as the places where the violets were grown changed, so did the way in which they were marketed. Thus when it was Verrières-le-Buisson, Fontenay, Sceaux etc. that were the centre of this cultivation, the growers made up uniform bunches for the wholesale trade. They were a kind of large bunch containing from 250 to 300 little violets with a few leaves around them.

The price they fetched was very variable; from 25 centimes in spring when they were abundant, to 4 francs in the winter. Today the grower hardly ever makes up these large bunches, and he himself makes small ones (a kind of buttonhole), which are sold for 5 to 25 francs the hundred; and when the latter price is exceeded he goes back to making large bunches.

This year, spring 1896, the very mild winter that we have just experienced has been favourable to the open-air cultivation of violets, and so the small bunches arrived in hundreds of thousands at the markets, coming from Marcoussis and the surrounding district. Each morning an entire train of six carriages came up to the Paris market. It is such an unexpected sight that the stranger who alights at Paris for the first time is amazed to see such a quantity of little carts full of violets making their way through the great city in all directions, leaving behind them a most sweet fragrance (a fair compensation for the smells of Paris). However it may be, these days when there is a glut of flowers, if they are pleasant for the great city and its consumers, are by no means so for the producer. I have seen bad days for the seller when prices fell to 3 francs a hundred, which was scarcely the cost of making up the little bunches.

Violets grown on a large scale are not usually mulched. However I should draw attention to a few growers who, by way of mulching, throw rotted-down manure on their fields of violets to help them through the middle of a dry year; but they are the exception in this.

CHAPTER XV

Attempts at cultivation with the use of the plough in the Paris area. Cultivation in the Midi and in provincial towns

Twenty years ago violets grown in the open were in such request that several cultivators tried using the plough; but these were rather experiments than a regular method of cultivation, for the difference consisted only in the following points. Instead of breaking up the ground by hand, it was done with a plough. Then the violets were planted in just three rows, put close to each other. The three rows did not extend more than 75 centimetres in width, and when the violets had made full growth the clumps touched and formed a vast carpet. The paths were wider, so as to allow the women who weeded and hoed to pass by when they were working. The middle of the beds could only be maintained by hand.

As can be seen, the work of the horse did not give any great advantage and, put briefly, it only served to maintain the paths and plough the ground; and this, with a plant that is difficult to reestablish, is not so good as digging.

In any case the services rendered by this work were not considered sufficiently remunerative, and after four or five years of fruitless trials this kind of cultivation was abandoned.

While speaking of cultivation I cannot however pass over in silence the methods used in the Midi, where Parma violets were first grown in France and at present one of the biggest centres of the violet industry, seriously threatening to ruin its rivals.

There is no special form of cultivation practised in the Midi. In far-off times violets were planted for six and eight years, the beds were carpeted with runners, and those flowers that appeared were picked. When the violets that were gathered for sending away as cut flowers became crowded together, they began to renew the plantations more frequently, so as to get larger flowers with longer stems. For just a few years now many have been growing violets in the way that has been described for Parisian cultivators, and many enterprises have been started by people coming from Paris. Some even, so as to have no check in the yield of flowers, begin to give them protection and put them in frames; without any heating, as this is not needed in their climate.

In all provincial towns of any importance in France violets are, generally speaking, grown quite well, because they are produced by florists wishing to have some flowers in winter. Some grow them simply in frames, others give them heat from the paths, according to their requirements. I should add that quite often these gardeners grow the two kinds: single sweet violets and Parma violets.

At Toulouse, where they grow very well, the latter are the subject of a large-scale trade for export. They are treated as at Paris; but, being grown in cold frames only, they yield splendid results. The flower without its stem is used by confectioners in their manufacture of crystallised violets. (see note at the end of the book).

I said earlier on that Northern Europe was dependent on France for violets. This is not the case with countries in the south. Spain, Italy and Turkey used not to cultivate them, or simply gathered a few that flowered in the wild. Today it is no longer like this. The cultivation of violets is highly regarded in these three countries; Paris exports plants of the most recent varieties to them in thousands, and these varieties grown according to the methods of the mother country are making rapid progress.

CHAPTER XVI

Cultivation of violets in North America: the United States

Speaking of progress, I cannot end this account of cultivation in these various countries without mentioning America, at least the United States.

Some twenty years ago violets were little if at all cultivated in the United States. Just a few horticulturists grew violets, not as a speciality but in glass-houses with other plants, kept at too high a temperature and in pots. Only scanty results were achieved. The most popular ones were the doubles, including the Parma violet 'Marie-Louise'. Single violets were so little regarded there that several horticulturists from America, who were travelling in France and to whom I offered some of our beautiful single violets, replied: "We would gladly take them, but we wouldn't sell them".

Today things are much changed. Some half-dozen enterprising growers, five of them French, have made this production highly esteemed and have popularised our beautiful varieties. Two of these cultivators do a large trade in violets. One of them, J.L.Desroches[168], has begun growing them on a large scale at San Francisco, but sales are still restricted for him by the lack of big centres. The other, who is more fortunate, is François Supiot[169] at Philadelphia. There he grows violets on a vast scale, and dispatches the flowers for sale to Washington, Baltimore, New York etc., where he finds a very profitable outlet. Contrary to cultivators down there in San Francisco, he only grows and sells single violets.

I cite these two growers because they have established the cultivation of violets on a large scale, especially from the point of view of cut flowers. Many other horticulturists are also engaged in this, but to sell retail only.

I should also say that the horticultural and agricultural societies, even those of the state, are by no means indifferent in this matter, and during the course of 1895 these various societies have exchanged a quite voluminous correspondence with me on this subject.

As for methods of cultivation, though they are similar to those in use in France, there are slight differences. Thus in San Francisco land is chosen where the temperatures are lowest and there is most cooling air, so as to avoid the worst effects of the summer, which is severe for violets. It is scarcely necessary to say that they are always grown and remain in the open, where the winter season is very favourable to them.

At Philadelphia the position is quite different. The winters there are very harsh, and so the plants must at least be given protection. During the last few years the Americans have begun to use frames for this.

I have no doubt that within a few years this cultivation will come to equal that of France, in view of the temperament of Americans. I should also do justice to these friends from overseas, who do not neglect anything that will enable them to achieve success. Almost every year, although at great expense, they travel to France so as to acquaint themselves with changes in methods of work, improvements and varieties that have come about in the course of time.

CHAPTER XVII

Growing violets in pots; how is it done?

Whilst surveying the various methods of cultivation, there are a few I must not fail to mention; one is that of growing violets in pots. If I persisted in saying that violets are not grown in pots, I would find many incredulous people who would say to me: "Where then do those thousands and thousands of pots of Parma violets that are sold every day in Paris and other great cities come from?" And there is the little secret; they are not grown in pots, they are put into pots when they are to be sold, and in this manner.

I mentioned when dealing with the cultivation of Parma violets that growing them was quite profitable, and that they were turned to account in two ways: firstly as cut flowers up until March; then secondly as pot plants. When the grower sees that his clumps are exhausted, that there are only enough flower-buds left to make the plants decorative, he chooses this moment to pot them up. With all good cultivators these clumps of Parma violets are well developed, neat, and have had their runners removed. As was said in the account of their cultivation, they are exceedingly suited for planting in pots.

For this purpose a large number of flowers are left to come into bloom. Then four clumps are put together, since Parma violets are never very vigorous, and planted in small rimmed pots 12 to 13 centimetres in diameter. All the pots of violets are put together in frames, where they are kept in a somewhat close atmosphere, in order to draw up the leaves and flowers that may have been crumpled or knocked down while the plants were being potted. In this way lovely pots of violets are obtained, which are usually sold about a week after they have been potted up. And for this reason they often cause disappointment to those who buy them, in not lasting very long in a room, and there are two reasons for this. The first is that they have not had sufficient time to recover in the pot; the second is that they rarely survive in the pot; they could not, they are too smothered and cramped there.

The grower has only one aim; to produce a lot of flowers that will make a fine show. The customer's eye must be seduced, everything depends on that; what happens afterwards is not his concern.

And so one often hears somebody say: "It's strange, we can't get such lovely pots of violets as are sold by the florists in Paris". That is true, but on the other hand, if you have only nine or ten flowers open at the same time instead of fifty to sixty, the plant survives much longer. On this subject I must go back a little. When I spoke of methods of cultivating violets, I declared that they were not grown in pots; this is a fact.

When it is a matter of obtaining practical results, normal yields of very fine flowers, you must have frames at your disposal and have a certain quantity of plants grown. But this is not the case with everyone. Many people have no frames but on the other hand possess a small cold or heated greenhouse in which there are plants for the winter; cyclamen, cineraria etc. Often they would like to have violets, even if only so as to have a few flowers from time to time to provide fragrance for the rooms.

Well, nothing is easier. We will raise some violets as everybody does, in the open in the garden, above all removing the runners. Then on the 15th October we pot them up, but not as they do when they are going to sell them. Only one good clump or two of medium size should be put in each pot.

When we have planted these violets in rimmed pots of 13 to 14 centimetres diameter, using good healthy soil, we take them to a place where they will be sheltered in bad weather, that is to say at the foot of a wall, in order to give them time to make a slow and sure recovery. If we are growing single violets we will not bring them back to our greenhouse until there are frosts. If they are Parma violets we will bring them in on 15th November.

We have to provide for the case where we have to do with a rather hot greenhouse; this is the most difficult situation. If possible the coolest, best ventilated place must be chosen, and they should be put very close to the glass. Unless these little observations are followed to the letter, our violets will immediately start into growth and produce tender leaves, and we can say goodbye to buds and blooms.

In a cold greenhouse, by always placing the pots close to

the glass and keeping to the same principles, there is a much greater chance of success, as the violets are in their element.

I have on many occasions seen, and have myself grown violets on a small scale in this manner, which have given satisfaction to their owners, but I have always noticed a discoloration of the flower. On the other hand the scent was more pronounced.

I also advise putting a few plants of 'Armandine Millet' in pots, especially in small pots, when it makes a delightful little flowering plant, as our illustration shows.

'Armandine Millet.'

CHAPTER XVIII

Harvesting seed; sowing; various methods of carrying this out

Gathering seed of violets is not difficult, but it is a good thing to make preparations. It is carried out at two seasons, in spring and in autumn.

In the autumn, just when the plants begin to form proper flowers, flowers which only become visible (1) at this moment, fertilisation takes place. In midsummer the organs of the flowers are so weak that they produce no seed. Harvesting seed capsules in the autumn can go on from 15th September to 15th November, according to the varieties.

(1) I say visible deliberately. I have often remarked that violets never stop flowering; during the summer they continually put out very short slender peduncles, barely visible, with the semblance of a corolla without calyx and petals.

In spring the same thing happens, but in the opposite order. The perfect flowers of winter are too impregnated with moisture, full of sap, and this, together with the cold, is prejudicial to fertilisation. It is only towards the end of the flowering season of each variety that, assisted by the spring temperature, seed is produced. This seed can be harvested from the end of March to the end of May. To do so successfully you must keep a careful watch on the maturity of the capsules, which, if not taken in time, will immediately hurl the seeds all round the mother plant.

When we have completed our harvesting, most of the capsules will have opened in the receptacles where we put them to dry. We have now only to sift them, in order to free them of other matter, and put them in reserve for up to four or five years. Of course the older our seed becomes the more it will lose its power of germination.

Few people grow violets from seed, apart from specialists, some amateurs and the occasional person who finds it easier to take away seeds rather than plants. In theory this is a good idea, but in practice it is not, and this is why. As three quarters of species or varieties, and especially the most beautiful, give little or no seed, this reduces the chances of obtaining any. Moreover the seeds of violets germinate with difficulty, even on the spot; all the more so when they have been dried during transport. Briefly, I always advise taking plants; it is much easier, they are neither very expensive nor bulky, and can be posted to the most distant places.

For those who wish to sow violets the task is not difficult. To obtain the best result you have only to let nature act on its own. Having gathered the seeds in autumn you sow them immediately in earthenware pots or pans in good, fine, sweet and fresh soil. The pots are put out in the open garden, with their rims at soil level. They are also covered with wire gauze or metal grating to protect the seeds from the invasion of voles and field mice. Unless this is done not a single seed will remain in the spring.

As we sow at the beginning of November, nothing else needs to be done; the pots remain moist enough until spring, the time when shoots, which have been developing all winter, start to appear. At this time you must not fail to keep them in a state of constant and regular humidity.

When the cotyledons are well up and the first little leaves develop, this is the time to prick them out in a nursery bed, in half-shade to begin with, and gradually put them where there is a good circulation of air. At the end of March, although small, they can be planted in their permanent place like large plants from divisions or runners. In the autumn they will be ahead of the others, since seedlings make a better recovery when transplanted and are much more vigorous than plants from any other form of propagation.

For those who do not have fresh seed available, have only seed they have bought or that has travelled far, and consequently the seed-coat has hardened, it will always be a good idea to stratify the seeds. For example they can be put in good fresh and moist composted material, the whole then going into a receptacle which will be kept moist and at a tepid temperature. After two months of this preparation the seed is sown in earthenware pots or other containers, towards the end of January if the weather permits, the ground being covered with cloches[170] and everything still kept moist. If the weather is cold or there is frost we can put our pots in frames or in a cold greenhouse, but the humidity must always be maintained.

The operation of sowing has two effects. The first is to regenerate the stocks, to make them vigorous. The second is the hope of obtaining interesting varieties. Unfortunately it is not easy to get good results, even in these two instances, and this is why. If you have common varieties of violets to regenerate, that is all right; you gather the seed in quantity, sow them without much trouble and you obtain plants almost identical with what you already possess, with the addition of new vigour and a mass of insignificant variations of your variety; in short, well and good. But the difficulty is in obtaining seed on the beautiful varieties. On these there are hardly any capsules, hardly any seeds; and then again, the few that you do gather are badly formed, they germinate poorly. Those that germinate are almost identical to the type of what is sown, quite often not as good, and it is a success when you obtain a worthwhile variety. In addition to these obstacles there are varieties which have never borne seed. They have existed for a very long time and have never varied. I hear someone say: "Why do you not pollinate them artificially?"

This is possible, it is true, but it is easier to say so than to carry it out. This is certainly the case with violets that produce little seed. These varieties do so only at the end of September or beginning of October. Well, as at least two months are required for the formation of these seeds, the flowers that formed them were in bloom about the end of July. At this time of the year the flowers of violets are barely visible, their organs malformed, the pollen difficult to assess, not easily transported. In a word, an operation carried out under such conditions is far from certain.

An unquestionable fact proves what I say. Let us take an example in the varieties that naturally produce a vast quantity of seed and sow themselves in the open ground, such as 'White Czar', 'Rawson's White', 'Wilson', 'Viola odorata rubra', 'Quatre-saisons odorante'. Here are five varieties that are lavish with their seed and are grown quite close to each other in a nursery without having shown any variation. It would be a good thing if artificial fertilisation[171] were easy to carry out. Had it been, then from my modest experience I think I should long ago have had an infinitely varied intermixture of these varieties; for our obliging assistants in such work, the flies, would not have failed to bring it about had it been possible, especially in view of the time of year – the months of July and August.

To sum up, if artificial pollination is not impossible, it is far from easy to carry out, and the result not at all certain.

CHAPTER XIX

Diseases and pests of violets

As is the case with all plants, violets have their natural and inopportune enemies. However, I must declare that the common enemy is not yet very dangerous and that with a little goodwill one can easily contend with all his inroads.

I say: as with all plants. It would seem that, according to me, everything we grow now is subject to all the disasters in the world. Everywhere you hear people say: "Oh! in the old days we didn't have these diseases, you can't get anything any longer, everything is sick, nothing grows any more". This is true in one sense; it cannot be denied that we see more diseases than a hundred years ago. But also I do not think that, a hundred years and more ago, people were seriously engaged in specialised cultivation, including the production of early vegetables and out of season flowers, leading to a remunerative trade, as is the case today. When a crop was unsatisfactory this falling away was attributed to disasters of various kinds; foul weather, or fogs, or frosts, or to all sorts of other causes.

Now things are no longer the same. The fields under cultivation are overworked; they are required to give the maximum yield each year without intermission, and people go about repeating the words of a few pessimists: "Nothing thrives any longer, everything is sick".

I do not believe that these words are well founded for, when these terrible prognostications are closely studied, quite the contrary emerges. Have such fine horticultural products as those of our time ever been seen before? Whichever way you turn there is fruit, flowers, vegetables, all in abundance, of good quality, large, inexpensive, and handsome into the bargain. I will not name any one product, since to be impartial I would have to name them all.

I prefer to come back to our violets and say to my readers: "How do you expect people a hundred years ago to have been able to record the diseases of violets, when they were scarcely cultivated, at any rate commercially, and no importance was attached to them? There is nothing surprising in the fact that, at that period, as in our time, mice might have nibbled leaves off woodland violets to make their nests; nothing surprising either that a parasite should have invaded violets growing in the woods. In short, all that may have existed then, but as there were no people financially involved, no close attention was paid to the matter".

Today things are different. Every year new fields are brought under cultivation, and at great expense. They are expected to cover costs and give a profit; and so the returns must always be ample. In this connection I must indicate the little disadvantages associated with the cultivation of violets. As I have said already, they are not very great, when dealt with promptly. Left to the march of time, they can on the contrary take on quite serious proportions and endanger a good part of our work.

CHAPTER XX

Concerning voles and field mice

"A tout seigneur tout honneur"[172] (Give every man his due). As mice and voles are the largest animals that cause damage to fields of violets, I must deal with them first. It is not that they do irreparable harm; however, if this rodent tribe[173] were left to establish itself in the area under cultivation you would have no reason to congratulate yourself. Here and there would be little burrows with heaps of earth around them, covering most of the plants; further off would be round nests made out of the leaves of violets. Whole frames full of violets would have had their leaves stripped off in a single night, and, an extraordinary thing, not one flower-bud nipped off (1). Well, granted that the buds will not have been nipped off, these heaps of earth, sometimes of droppings, these gnawed leaves piled up together, all this will not constitute cultivated land in perfect condition.

(1) Regarding these leaves that have been cut off, I have this year made another fairly bizarre observation. As I said, I had seen leaves that had been cut off without a single flower-bud among them, and that when the plants were covered with buds. Alongside I had a greenhouse full of cyclamen[174] of medium size, full of bud. One morning I was cast into despair on seeing that a large part of my cyclamen flower-buds had been nipped off, although the leaves had not been touched. This was the opposite of what happened to the violets. The next morning I had caught five voles.

The only thing to do is to get rid of these rodents, and nothing is easier. A few mouse-traps with three entrances, baited with corn or flour, will relieve you of them. If that does not suffice, set up the gardener's mouse-trap; it is infallible and very cheap. All that needs to be done is to stand a bowl 10 or 12 centimetres in diameter upside down on a tile, saucer or any other flat surface and raise it sufficiently to place a walnut under the rim. The nut, that has been prepared by breaking one side, is placed with the broken part facing into the bowl, and the trick is done. The rodent, aware that there is one part of the nut more convenient to attack than the other, goes round under the bowl and sets about the nut, which, as it is lightly balanced under the rim, slips away, and the rodent is held captive underneath the bowl.

However, I must say that an invasion of this kind is not common. So much the better; those who are not attacked by the problem are saved both time and trouble.[175]

CHAPTER XXI

Aphis: green aphis; brown aphis in the soil; types of the aphis family

Unfortunately violets are no more free from aphis than are other flowers. They are not attacked every year, as are roses for example; but although an infestation is not a regular thing, it is none the less unpleasant.

I begin with the green aphis, the one which lives on the leaves and flowers of the violet. It attaches itself right along the stems, extracting their sap and twisting them in such a manner that you often see a violet acting as the mouthpiece of a green hunting-horn. All parts of stems that have been sucked are twisted and deformed, flowers are distorted and spoiled. There is a radical cure, that is by spraying with tobacco. This never fails, but it is awkward using it with violets, for they are principally grown for their scent, and the nicotine that is used always leaves a certain smell of tobacco. I would rather advise that, if the attack is noticed at its first appearance, powdered tobacco should be scattered over the plants. This does not give off any smell, and is sufficient to stop the aphis attack. If it is noticed when the violets are well infested, two serious sprayings will rid you of the pest. However, for some time the violets will smell of tobacco, and to get rid of this it will be useful to spray them thoroughly on two occasions with water. But fortunately such invasions do not often occur with violets.

Another kind of aphis attacks the roots and main body of the plants. This aphis is brown and more voracious than the preceding one. I think it is the same as that which attacks the sow-thistle[176]. For a long time I did not notice that this disorder of violets was caused by an aphis. This is what it does. In August, when the violets are growing and beginning to burgeon, you suddenly see some clumps among them that seemed to be in full health wilt and die in a few days. For a long time I was not disturbed, but on examining the dead clumps I found that all the herbaceous tissues had been devoured, leaving only the woody core. This state of things intrigued me, I studied to find out the cause of this devastation. I kept watch and discovered some little creatures. They attack the basal stem, the "trunk", all below ground, devouring, as I said before, the herbaceous tissues of the central rooting parts of the plants. They leave nothing but the wood, and this determines the plant's death.

They usually depart before the violets wilt.

I must draw attention to another insect, much smaller than the aphis, but much more terrible. I refer to the red spider, commonly known as grey spider, which is a mite (of the family Arachnida).

CHAPTER XXII

Red spider, commonly known as grey spider

This spider, that is only too well known to gardeners, attacks violets on the under surface of the leaves, especially in August. It causes the leaves to turn yellow, and finally reduces the plant to a deplorable state of anaemia. However, the violet is not its favourite host; French beans, the tuberose, strawberries, the elms and lime trees of our towns are annually infested with them. More often than not it is these same plants that transmit the red spider to violets, and up to now no radical treatment has been found that will get rid of them.

The best method, when it is feasible and the plants will stand it, is to keep them in a constant state of humidity; this is what the spider dislikes most of all, and will often force him to let go. It often happens naturally in autumn, when great heat is followed by the rains of September and October. Then the leaves at the centre of the plant become green again, and those all round the plant that has been badly affected turn yellow and die. It is a good thing, if you can, to remove these wretched leaves, so as to hasten the departure of the spider.

CHAPTER XXIII

Cryptogamous diseases

Having more or less acquainted readers with the creatures harmful to this kind of plant, it remains for me to say a few words about fungi that attack violets: Peronosporales, mildew etc.

The first instance, which is known to all violet growers, is a fungus with very large fibres that develop very rapidly and form tillers of mildew, especially where leaves are crowded together. This fungus is not due to anything abnormal in the plant, it is the result of great humidity within the frames during the winter. Under the same conditions all the plants are attacked by it; and so preventive treatment is always the best. Good ventilation during fine weather, plants that are in a healthy state and have been well prepared will prevent the appearance of this drawback or disease; for this Thallophyte has always existed and has always developed on plants that are sickly and insufficiently ventilated rather than on those in good condition.

The second parasite, from the family Peronosporales, is more recent; I should say has become known more recently. It was first made known to me by one of my good clients in America. As with mildew on vines, it breaks out suddenly, and anyone who does not know this disease is surprised by the speed at which it progresses.

Indeed it starts in a bizarre fashion. You would say that a small pellet of hot lead, having burnt the leaf, had formed a brownish-black spot. The spot grows in size, carrying with it a bluish circle, which is itself surrounded by another yellowish circle, the result of discoloration of the leaf. If the weather is favourable for development the circles grow larger, invade the whole leaf, and at the slightest burst of sunshine it becomes deadened as though it had been burnt and reduced to ashes.

Some plants die from the attack of this disease. However I must add that when I first heard of it it seemed to me more dangerous than it is. For several years now, I do not know whether I owe it to the sulphur I spread over my plants, the fact remains that they are in good condition.

I do not claim that sulphur is a cure for this disease. No, that is far from my thought; it is just that I consider that it serves to isolate the leaves from the terrible humidity of the long winter months. To do this the sulphur must be well distributed and in the form of a fine, almost invisible powder.

Finally, to end this chapter I shall pass on to the third cryptogamous disease.

This last disease is fairly serious in one sense, that is to say that up to the present it has only attacked two or three varieties. This is fortunate, since for quite a long time it has deprived the florist-gardeners of Chevreuse of one branch of cultivation.

These growers, great cultivators of pansies, daisies etc., used to deliver to the Paris market double white and double blue violets by the thousand little wicker trays each market day. Today the customer has difficulty in lighting upon a few small specimens.

As I wanted to find out what had led the growers to abandon this plant, I cross-examined several of them. They all gave the same reply: "Since the violets were no longer growing vigorously or coming to anything, we gave them up".

I myself narrowly escaped losing these two double violets. But it was in particular 'Blanche Double de Chevreuse' and 'Belle de Châtenay', which is also double, that were the worst affected.

As I wished to understand the disease that was decimating the plants like this, I made a study of the affected plants. And it was worth the trouble. All my clumps of violets were full of swellings; stems, leaves, flowers, even the runners, everything was in the same state. The leaf stems, which should have been the thickness of a small matchstick, were as big as a little finger, and there was a swelling in every place. I could not compare these swellings better than with those that the woolly aphis makes on apple trees, and for a long time I even thought they were the effect of a similar aphis. Nothing of the kind. Despite all my research, although I was assisted by quite powerful instruments, I was unable to ascertain the presence of an aphis; all that I found was a fungus. And my observations have led me to record in all its detail the progress of this fungus, which is very curious to watch.

How does it originate? I cannot say for certain (1). The first trace that I have noticed on a plant is a small longitudinal filament quite like a thread. Soon little hard grey points are formed; from there other little grey filaments emerge which, acting like an octopus, clasp firmly and smother the stem that is attacked. The part that remains free swells even more from this, and forms a longitudinal blister that sometimes covers the place that is being attacked. When they reach this phase of development the tumescences burst lengthways, revealing a blackish-chocolate coloured powder which, at the least breath of air, blows off from every cavity on to the leaves.

(1) This year, quite recently, I learned what is almost certainly the origin and method of propagation of this cryptogam. From my modest observations I consider it to be entirely due to a black powder coming from the fungus when it has matured and which, spreading over all the veins of the plant, inserts into them cryptogamic filaments; these, finding a suitable medium for their development in certain varieties, grow there rapidly.

It is fortunate that only the variety 'Blanche Double de Chevreuse' is favourable to this disease.

I should add that for many years I had recorded these excrescences on the violet; but whether because the varieties then known were not subject to contamination, or for quite a different reason, it did not cause any devastation and these swellings were regarded as accidents of nature.

This is the furthest point of development of the disease. I have seen certain plants affected in all their vital parts at the same time. Then all these swellings become mildewed and rot, so bringing about the death of the plant.

For several years I was greatly intrigued by this disease. I planted violets out in very good conditions so as to get my plants growing well and make them strong, in order to resist the scourge. Nothing had any effect. On the contrary, the stronger and fresher the plants were, the more forcefully the disorder developed. Feeling my way all the time, I removed all the diseased parts and plunged the plants completely in a bath of Bordeaux mixture. They were then taken out and treated with flowers of sulphur. This was perhaps a little drastic. In any case it worked, and I had the pleasure of seeing my plants making headway and reaching the autumn in a very healthy state and flourishing[177]. Since then, when a few slight traces of a recurrence of the malady appear, I either remove the sick plants or else I treat them as I have described above, and I obtain irreproachable specimens.

I will end with a disease which is not one at all. It is of blind violets, to use the technical term, that I wish to speak; that is to say violets with crowns that are not well formed. This is more often than not due to the position they occupied while being raised, either because they were too close to each other or because they were in the shade. In both cases it is always the lack of air that is the cause. The leaves have developed too rapidly, have made the plant weakly and unproductive. In this case nothing can be done. We must bring back our violets to shelter or under glass and get rid of the plants formed like that. Blind violets are very easily recognised among others, as the leaves emerge all at once instead of being alternate. At other times in the semblance of a crown you see five or six little flower-buds, all at the same height instead of being successive.

Here ends my work. In carrying it out I have not for a single moment deviated from the modesty that is called for by my subject. In presenting this book to my readers I do not forget the ladies among them, to whom I am happy to offer it together with these four lines of verse with which a poet of the seventeenth century composed a madrigal about these two sisters, the violet and the lady:

La Violette

> Modeste est ma couleur, modeste est mon séjour,
> Franche d'ambition, je me cache sous l'herbe;
> Mais si, sur votre front je me puis voir un jour,
> La plus humble des fleurs sera la plus superbe.
>
> (Desmarets de Saint-Sorlin[178])

(Unpretentious is my colour, as is my dwelling;
Devoid of all ambition, I lie concealed beneath the grass.
But if some day I find myself adorning your brow,
Then the lowliest of flowers will be the proudest of all.)

NOTES

On the manufacture of quality sweets made of whole violets

In connection with the violets of Toulouse and several other centres (exclusively double Parma violets), I would not wish to pass over in silence an industry, which has only been in existence a few years, and that is developing very well and growing daily.

In the South of France violets have always been used in perfumery (essential oil of violets and other products), but the idea had never occurred of making sweets from them: not just sweets scented of violets, but made entirely of high quality violet flowers. Today this idea has turned into an established business, and at present Toulouse, Nice, and even Paris, trade in this sweet.

This manufacture, which began several years ago with the investment of a few thousand francs, has now in 1896 reached considerable sums and promises to expand greatly. And yet this business is based entirely on a sweet of irregular shape made out of a whole Parma violet flower. The violets, with the stem removed, are passed through thick syrup, which impregnates all the petals and thus gives body to the flower without taking away its shape, colour and scent. The violets are, to use the technical term, crystallised[179].

Briefly, here are perfect sweets, pretty and of exquisite flavour. Despite having been processed, the violets retain their mauvish-blue colour and the refinement of their form. As I said above, each year witnesses the growth of this industry. In Toulouse five important firms are seriously engaged in it. The firm of Arnould[180], of rue Saint-Étienne, has the greatest output of this manufacture; it was, moreover, the first firm to produce these sweets. There are also several businesses engaged in this trade in Paris and Nice.

Between 500 and 1,000 Parma violets are needed to make a kilo of sweets, according to the size of the flowers. The price is also variable, depending on whether the flowers are large and well grown, or small. The price per kilo is from 6 to 9 francs.

Although part of the output is consumed in France, the main outlets for this product are in North America: New York, Chicago, Philadelphia, and even San Francisco. Some colonial territories also buy a fair amount.

And so our growers are delighted with this new use which, in good time, will stimulate the sale of this flower, bringing them good profits.

A new variety not previously described

While this work was being prepared a new variety, with a yellow flower, appeared. I say new, although it is not really new, since it was gathered by a postman in the department of Indre, where it was growing wild on the edge of a wood.

It is distributed commercially under the name of Viola odorata sulphurea[181] (odorata is questionable). However that may be, this is a novel plant, almost beautiful, a unique colour in its family, citron yellow, chamois at the throat, very pleasing. The plant seems to be sturdy and vigorous, with fine deep green foliage, and extremely floriferous, especially in spring. This little violet is destined to provide an agreeable variation in our collections and will, I hope, establish a place for itself among the other violets of our time.

CONTENTS

	page
Dedication	32
Introduction	33

First Part

Historical account of violets; the author's observations	35
Legends about, and works on the violet	36
Kinds of violets spoken of in 1566	38
Violets are cultivated in 1690. Some notes on this cultivation by de la Quintinye	40
Varieties grown at the beginning of the eighteenth century; development in the South of France	44
Violets in Haute-Garonne, and a few words on Angoulême	49
1700 to 1750. Violet cultivation in the Paris area; how it began	50
1750 to 1780	51
How violets were sold in Paris in 1780 and the following years	51
1795 to 1825. Cultivation draws nearer to Paris and spreads	52
1825 to 1835. Improvement and considerable extension of cultivation	53
1835 to 1843. Trading is regularised; violets sold wholesale	54
1843 to 1859. Violet cultivation, especially in the open, is all the rage	55
1859 to 1870. Two excellent varieties make their appearance; prices obtained	56
1870 to 1880. An avalanche of novelties: 'Czar', 'Millet père', 'Gloire de Bourg-la-Reine'	58
The last period, 1880 to 1896; fine new varieties	62
Concerning the influence of resin on the colouring of leaves	66

	page
A few notes on 'Princesse de Galles'	67
An account of Cucullata violets	72
Kinds of violets: violets with stems and without stems at the same time	74

Part 2
Cultivation of Violets in Woodland, Gardens, Frames and Glass-houses

Ch. I Cultivation in bygone times	78
Ch. II Planting in woodland	84
Ch. III Growing violets in small gardens	86
Ch. IV Varieties suitable for small gardens	87
Ch. V Violets in large gardens and on great estates. What can be expected of them and which varieties to use	88
Ch. VI Parma violets	91
Ch. VII Late-flowering violets – Viola cucullata grandiflora and other so-called tuberous violets	94
Ch. VIII Methods of cultivation; preparation of plants	96
Ch. IX Cultivation of Parma violets and hardy double violets. Preparation of these divers plants	100
Ch. X Propagating and cultivating Viola cucullata	103
Ch. XI Single sweet violets in the winter season; method of forcing; the attention they require	105
Ch. XII Forcing Parma violets; the results that can be obtained from this	109
Ch. XIII The forcing of Parma violets and single violets, as practised by specialists in the Paris region	112
Ch. XIV Violets grown in the open; their cultivation and development	115
Ch. XV Attempts at cultivation with the use of the plough in the Paris area. Cultivation in the Midi and in provincial towns	118
Ch. XVI Cultivation of violets in North America: the United States	120
Ch. XVII Growing violets in pots; how is it done?	122
Ch. XVIII Harvesting seed; sowing; various methods of carrying this out	125
Ch. XIX Diseases and pests of violets	129
Ch. XX Concerning voles and field mice	131

Contents

Ch. XXI Aphis; green aphis; brown aphis in the soil; types of the aphis family................ 133
Ch. XXII Red spider, commonly known as grey spider . 135
Ch. XXIII Cryptogamous diseases 136

Notes On the manufacture of quality sweets made of whole violets 140
A new variety not previously described 141

TRANSLATOR'S NOTES

In these notes M. precedes words from the text of *Les Violettes* and D. precedes words from the text of Dodoens' *Florum et coronariarum*...

All references in the translation and elsewhere in this book to the Society, Horticultural Society, Paris Horticultural Society etc. are to the French National Horticultural Society (Central or Central and National before 1880). Its journal was *Le Journal de la Société Nationale d'Horticulture de France* (*JSNHF*) (*JSCHF* before 1880).

Cleistogamy is "the production of flowers that do not open to expose the reproductive organs, so preventing cross pollination" (*Penguin Dictionary of Botany*). These are the invisible or unseen flowers referred to by Millet, and most single cultivars of violets produce an abundance of seed from such flowers. This seed has a substance attached to it that is attractive to ants, and they drag seeds away. A collection of violets may soon become muddled, and seedlings of unexpected complexion appear where least expected.

When in early April 1898 a copy of Millet's *Les Violettes* was received by the Horticultural Society for its library, the book was immediately forwarded to the committee of the Joubert de l'Hiberderie Prize for consideration. (*JSNHF* 1898 p.350). This award was for a work, already published or in manuscript, on a horticultural subject treated in a practical way. It was founded by a bequest of Dr.Joubert de l'Hiberderie, who died in 1889, and the prize could amount to 2,500 francs and be awarded annually. (*JSNHF* 1893 p.713). At the end of April 1898 300 francs was awarded to a schoolteacher named Nicolas for a work called *Veillées horticoles*. (*JSNHF* 1898 p.579). The next year the full amount of 2,500 francs was given to M.Villard for an illustrated book, *Les Fleurs au XIXe siècle*. (*JSNHF* 1899 p.645)

The illustrations. Millet thought that by photographing the violets himself and having them engraved for printing he would ensure an accurate representation of each of these varieties. In reply to my enquiry at the Bibliothèque Nationale I was told they did not consider there was anything unusual in the method employed. Ironically his full page illustration of 'La France' (fig.10 in the book) was reprinted in *Le Petit Jardin* (1906 p.323) entitled 'Princesse de Galles'!

1. M.Doin. Ocatave Doin was a publisher of books on horticulture and agriculture. Millet's *Les Violettes* and his book on strawberries *Les Fraisiers* came out as part of a *Bibliothèque d'horticulture et de jardinage* in 1898. When Millet was writing *Les Violettes* Doin was President of the Orchid Committee of the SNHF and exhibiting at the Paris shows. He had an important collection of orchids at his house, the Château de Semont, near Dourdan (Essonne). He died in December 1919.

2. article.....1878 This was *Notes sur les différentes cultures de violettes aux environs de Paris* (JSCHF 1878 pp.230 – 237)

3. 1566 It is difficult to explain Millet's dating of Dodoens' work. The article on the violet that is translated here, and the accompanying illustration, were first published by Plantin in *Florum et coronariarum odoratarumque nonnullarum herbarum historia* in 1568. A second edition was printed in 1569. In 1583 the same article, with very minor alterations and the addition of the illustrations of a double violet and an unscented 'dog' violet, appeared in Dodoens' collected works, *Stirpium historiae pemptades sex, sive libri XXX*. This too was published by Plantin. The wood engravings were designed by Peeter van der Borcht; the blocks were cut by Arnold Nicolai and Gerard Janssen van Kampen. This engraving was later used in other herbals, and finally appeared in Johnson's edition of Gerard's *Herbal*, which was based on a translation of Dodoens' work.

4. la Quintinye 1690-1730 These dates are of editions of his works. The title of the first edition of 1690 is *Instructions pour les jardins fruitiers et potagers*, hence Millet's "instructions on gardening" (p.40). Jean de la Quintinye (1626-1688) was "Directeur des Jardins fruitiers et potagers du Roy", he was the greatest authority on the cultivation of fruit and vegetables of his time. After his death his writings were published, and later translated into English by Evelyn and also by London and Wise.

Violets appear briefly in the section on the vegetable garden, as they were grown for ornament as well as for use in salads. In editions from 1695 an anonymous work on flowers was added to make this authoritative book more complete. This was not taken very seriously (it does not appear in English translations) and in the 1730 edition a full explanation that la Quintinye did not write it, and had never written about flowers, was printed facing the first page of this section (Vol.2, p.358)

5. very fine M. grand et beau

6. 1898 The introduction was written just before publication. As he mentions in the book, the work had been composed during 1895 and 1896. The introduction to his other book in the series, *Les Fraisiers*, is dated 15th October 1897. As they were both published early in 1898 it looks as though he had to present the manuscript on Violets without being able to revise it.

7. Napoleon I (1769-1821) In France violets came to be associated with Napoleon by his supporters, especially from the legend that he had said he would return from his exile on Elba "with the violets". It continued to be the emblem of the Napoleonic faction throughout the nineteenth century. Millet, as a commercial grower of violets, had no objection to anything that was good for trade. (see also note 84)

8. These paragraphs are translated from the same source as the rest of this section; they come in the middle of Dodoens' article on the "black or purple violet".

9. Theophrastus. A Greek philosopher, pupil of Aristotle; he died in 287 B.C. His *Enquiry into Plants* was translated into English by Sir Arthur Hort.

10. purple, looking almost black M. à cause de sa couleur, rouge tirant sur le noir D. a nigricantis purpurae florum colore

11. Pliny C.Plinius Secundus, known as Pliny the Elder, was born in 23 A.D. in Northern Italy. He died in 79 as a result of his attempts to observe the effects of the eruption of Vesuvius, and to save lives. His massive *Historia Naturalis* is divided into thirty-seven books.

12. species of similar colour M. des espèces approchantes par leur couleur. This should be "to distinguish it from other violets", that is from other plants that were then called viola (D. a caeteris Violis discerni, as in Pliny). "*Viola*. The Latin name for various sweet-scented flowers, such as violets,

stocks, wallflowers, and derived from the same source as the Greek 'ion', which in its earlier form had an initial letter corresponding to v or w, the digamma, which was later lost." (Smith and Stearn)

13. Nicander M. Dans sa description de la terre Nicoder D. in Geoponicis Nicander... Geoponici or Georgici were books on agriculture (cf. Virgil's *Georgics*). Nicander was a Greek poet, priest of Apollo and physician in the 2nd c. B.C. Apart from two poems, on poisons and the treatment of envenomed wounds, his verbose works have almost entirely disappeared. But Athenaeus quoted some fragments, including: "Nicander in the second book of his work *On Farming* ... says that the ion was conferred first on Ion by certain Ioniad nymphs". (Loeb edn. p.149)

14. Hermolaüs Hermolaüs Barbarus was a 15th c. scholar who edited works by classical writers. Haller lists an edition of Pliny's *Natural History* printed in Rome in 1470, with emendations by Hermolaüs. He was presumably responsible for emending Ion to Jupiter (Jove) in the passage from Nicander quoted by Athenaeus.

15. that sprang up from beneath her feet M. la terre produisit sous ses pieds des fleurs poussées à cette occasion D. terra florem hunc pabulo eius produxerit – produced this flower as food (or fodder) for her. Perhaps the translator thought the reference to food indelicate at this point.

16. same meaning as that for heifer M. même signification de celui de génisse D. Atque inde etiam Latinis Viola, quasi vitula extrita lettera t, dici putatur. Gerard's version is: "and thereupon it is thought that the Latines also called it viola, as though they should say Vitula, by blocking out the letter t". The Latin word vitula means a calf or heifer.

17. Servius A Latin grammarian of about the 4th c. A.D., who was known for an elaborate commentary on Virgil.

18. vaccinium M. vaccin D. vaccinium I suppose vaccin is a popular name for the plant, as Millet uses the word. "Vaccinium. Blueberry, bilberry, cranberry. A Latin name apparently derived from the same prehistoric Mediterranean language as the Greek hyakinthos and transferred to these berry-bearing shrubs". (Smith and Stearn). This explains the strange alternatives given for this word in translations from the Latin.

19. The white privets fall... M. has two translations,

printed: Les blancs troènes tombent, on cueille les airelles noires (Les blanches clématites tombent, les noires violettes sont cueillies) D. does not, of course, give a translation.

20. However Virgil... M. Cependant Virgile, dans son églogue Xe, montre que la violette (viola) diffère de violarii vaccinium (il y a des violettes noires, violiers) D. Virgilius tamen, Ecloga 10 vaccinium a viola differre ostendit. Et nigrae violae sunt, et vaccinia nigra. I presume Millet inserted violarii and violiers to show that he thought vaccinium here referred to a flower and not a whortleberry. M. omits the line of Virgil, and so makes his explanation ambiguous.

21. Vitruvius M.Vitruvius Pollio served under Julius Caesar in the African war, 46 B.C. His treatise on architecture was written in old age and dedicated to Augustus. Dodoens assumed that Viola here referred to *Viola odorata*, but it clearly means a "violet" with a yellow flower, perhaps wallflower; hence the translation in the Loeb edition "yellow violet". The confusion by Dodoens may stem from the account in Pliny, first describing a vegetable dye for imitation of sil (ochre) and then a dye from dried petals of violets for a substitute or adulterant of blue.

22. whortleberry M. vaccin D. Violam a Vaccinio distinguit

23. red or Attic ochre M. rouge (ou sil attique) D. Silis Attici colorem "Red" seems at best superfluous. Sil is ochre. In his translation of Pliny, bk.33, Philemon Holland has: "Polygnotus and Mycon were the first painters who wrought with Sil or Ochre, but they used only that of Athens in their pictures". Hence Attic ochre.

24. dyers M. teinturiers D. tinctores In Vitruvius it apparently means specifically stucco painters.

25. Eretrian earth M. érétrie D. eretriam Vitruvius has cretam. The descriptions in Vitruvius and Pliny are very similar. It took its name from the territory that produced it, and came in two forms, one white and the other of an ashy colour. Bailey gives: Eretria terra, or creta ?magnesite ($MgCO_3$). pouring Eretrian earth... M. et la répandant sur de l'érétrie, which is a mistaken translation of Dodoens (and Vitruvius).

26. whortleberries M. violier D. and Vitruvius vaccinium. Violier occurs in la Quintinye with the meaning "violet plant", but I am not sure why Millet uses it here.

27. M. <u>Rembert de Dodone, le Dr.Rembert.</u> Rembert Dodoens, latinised as Rembertus Dodonaeus, was born in 1517 in the Spanish Netherlands at Mechlin (now Malines) and died in 1585. Like other great herbalists of the 16th c., including our William Turner, he was an admirable scholar and humane doctor. He had the misfortune to lose his possessions when Mechlin was sacked by the Spanish troops. He was personal physician to the Holy Roman Emperor from 1574 to 1578. For a fine prtrait of Dodoens at the age of 35 see Anderson's *Illustrated History of Herbals*.

28. <u>1564</u> The only explanation I can think of for this date is that Millet was allowing two years for the preparation of the woodcut. To compound the error the printer gave the date beneath the illustration as MCLXIV.

29. <u>fruits</u> *Florum et coronariarum...* does not of course include fruit. One book of *Stirpium historiae...* is devoted to fruit trees, well illustrated, but it does not include garden varieties.

30. In the 1569 edition of *Florum et coronariarium...* (I have also seen the 1568 edition) the names are given for German, French, Belgian, but not Italian. In 1583 Dodoens added Italian, Spanish, English and Bohemian to his list.

31. <u>first... double violets</u> Not strictly true; for example Gesner in 1551 and 1561; and Dodoens himself in his *Cruydebook* (1561). Millet is not consistent in his use of the term "double violet". He usually means hardy double violets unless referring specifically to Parma violets.

32. <u>double blue and double pink</u> Millet says elsewhere that the double blue never bears seed, and this seems to be the usual experience today. From illustrations in the herbals it is clear though that many double violets grown then were quite different from those of later times. Some had distinct spurs. The words "both purple and white" were omitted from the translation of Dodoens in M., which accounts for Millet's perplexity.

33. <u>thinner, rounder</u> M. <u>plus minces, plus ou moins arrondies</u> D. <u>rotundiora, tenuiora</u>

34. <u>slender stems</u> M. <u>légers stylets</u> D. <u>tenues styli</u> A French translation of the time has "queues".

35. <u>deep purplish-blue</u> M. <u>d'un rouge bleu foncé</u> D.<u>colore in caeruleo subnigro purpurei</u>

36. <u>sides and lungs</u> M. <u>des reins et des poumons</u> D. <u>laterum et pulmonis.</u> Gerard has "sides and lungs".

37. <u>drosaton or serapion</u> M. <u>gâteau ou separion (sic)</u> D. <u>drosaton sive uti Actuarius nominat serapion</u> (drosaton and serapion in Greek). Millet's intended meaning of "gâteau" is not clear; some liquid preparation must be referred to, and "drosaton" be derived from the Greek for dew, as in the genus *Drosera*. In Costaeus both the man Serapion (a medieval authority referred to alongside Avicenna) and serapion (a medicament presumably named after him) are mentioned.

38. <u>Actuarius</u> According to Haller he lived in the 13th c. or early 14th c. and wrote on the composition of medicines (purgatives etc.). He took much of his information from Galen, but also supplemented Greek knowledge with that of the Arabs.

39. <u>ejected</u> M.<u>injecter</u>, which sounds painful. D. <u>educit</u>

40. <u>are taken</u> M. <u>quand on mange</u> D. <u>sumptum</u>

41. <u>unripe olives...omotribo</u> M. <u>olives fraîches...amotribo</u> D. <u>omoterbes or omfakinon</u> (all in Greek). "Omphacinum" occurs in Latin script in Costaeus: "Oleum, sit omphacinum, aut amygdalinum, ex quo oleum violatum fiet". (The oil from which oil of violets is made may be from unripe olives or almonds).

42. <u>Messues</u> He is a much mentioned authority in 16th c. herbals. Costaeus wrote a massive commentary on his works. Three books on medicine, in Arabic, are attributed to Messue the Younger, who is said by Leo Africanus to have lived 926-1016 A.D. If these dates are correct they seem to confirm his medical skill. He is said by Anderson to have been "a Jacobite Christian of probable Greek origin".

43. <u>are not only less cooling but also appear to have acquired some heating power</u> M. <u>rafraîchissent moins bien si elles ne donnent de l'inflammation</u> D. <u>et minus refrigerant, et caliditatis nonnihil adeptae videntur</u>

44. <u>Galenus</u> He was born at Pergamum, Asia Minor, c.130 A.D. and died c.200 A.D. Galen was the most influential writer on medicine among the ancients and his works were translated into Arabic, Hebrew and Syriac. He also wrote on other subjects, including logic, grammar, ethics, philosophy and literature. He became surgeon to the gladiators at Pergamum to increase his knowledge of anatomy and surgery, and later became physician to the Emperors Marcus Aurelius and Severus. His accurate observation caused his work to be standard until the 16th c.

Translator's notes

45. <u>Dioscorides</u> A Greek physician. He was active about 60 A.D., when his only extant work, *De Materia Medica*, was probably written. The oldest known manuscript is in the *Juliana Anicia Codex* of 512 A.D. This contains illustrations said to be modelled on those of Cratevas, a Greek herbalist and artist of the first century B.C. Dioscorides remained the great authority through to the Renaissance. The herbalists attempted to identify all the plants he recorded, and this led to realistic portrayal and examination of all the plants that were studied, and eventually to the development of modern botany and botanic painting.

46. M. omits a three line paragraph in Dodoens that gives further information about the treatment of this distressing complaint, "the breaking out of the fundament".

47. <u>migraine</u> This is the perhaps euphemistic word in M. D. has <u>crapula</u> – drunkenness.

48. <u>drives away scorpions</u> D. <u>semen scorpionibus adversari</u>. The idea was that scorpions would be repelled by something in the seeds and you wouldn't be infested by them. Those who slept rough in North Africa during the 1939-1945 war will know how useful this would be. However, it is a corruption over the ages from Pliny's: "The seed of the violet neutralises the sting of scorpions".

49. <u>Aster atticus</u> This is figured and described by Dodoens in a rather broad way to include a form with yellow flowers. The reference is probably to *Aster amellus*. Dodoens quotes from Virgil's *Georgics* Bk.4, 1.271-

> est etiam flos in pratis, cui nomen amello
> fecere agricolae, facilis quaerentibus herba;
> namque uno ingentem tollit de caespite silvam,
> aurens ipse, sed in foliis, quae plurima circum
> funduntur, violae sublucem purpura nigrae

(A flower too there is in the meadows, which farmers have called amellus, a plant easy for searchers to find, for from a single clump it lifts a vast growth. Golden is the disk, but in the petals, streaming profusely round, there is a crimson gleam amid the dark violet) (Loeb trans.)

50. <u>violets... not cultivated</u> Millet confines himself to the history of violets in France. They were grown in ancient times in the Middle East, in Greece and in Rome. In 1576 Lobel recorded double violets being used for making violet syrup at Antwerp. A warning that double violets may revert to single

flowers if they are not frequently transplanted, and lists of various colour forms of violets show that they were much grown in gardens in the sixteenth and seventeenth centuries. One interesting branch of the cultivation of violets was mentioned by Martyn in 1807. He gave the dosage of violet syrup for children, then added: "This syrup is very useful in chemistry, to detect an acid or an alkali: the former changing the blue colour to red, the latter to a green. For this purpose Violets are cultivated in large quantity at Stratford-upon-Avon".

51. tree violet Despite the description given: upward growth of shoots; milky substance in the roots that can be broken up for propagation; scent like storax; the appearance of the flower, Millet persisted in the unlikely identification of this fabulous plant with the hardy double violet 'Patrie' (syn. 'en arbre'), and in attributing the article to la Quintinye. In 1901, in the *Revue Horticole*, Mottet pointed out the discrepancy.

52. origin of the Parma violet The tradition in Italy, at Naples where it seems to have been known first, was that it was brought there by the Bourbons from Portugal (Portugal was annexed to Spain from 1580 until 1640). Hence the names Neapolitan Violet and Portuguese Violet. From Naples it was taken to Parma; and from Parma, apparently by Count Brazza at the beginning of the 19th c., to Udine. It presumably came to western Europe from the Middle East, North Africa or Turkey. The Parma violet is included in *Le Bon Jardinier* from 1805; a pale semi-double flower. It must be a very ancient plant, always valued for its long period of flowering and almost excessive production of strongly scented flowers.

53. double... wild Botanists would not be expected to record double flowers, however much they might be sought after by gardeners. Two references are: "It has been found wild with double flowers" (Miller's *Dictionary of Gardening* ed. Martyn, 1807); and "doubles, blue and white, found wild and easily propagated by division" (Phillips, *Flora Historica*, 1824). Among the many illustrations of double violets in herbals and florilegiums of the 16th c. and 17th c. there is a wide range of forms, from those with long spurs like single violets to those without a visible spur and with symmetrically arranged, tightly packed petals resembling a double hepatica. The earlier double violets that had been so

popular in England and were admired by Cobbett on his rides seem to have been abandoned by the mid-19th c. in favour of more regular blooms acceptable to the florists, with their exacting rules. For an account of the various forms of double flowers found among violets see *Vegetable Teratology* (Masters, 1869)

54. become double... cultivation Ragonot-Godefroy wrote in 1844: "d'autres se sont doublées par la culture" (other (violets) have become double as a result of cultivation). Millet was referring to observations of this kind.

55. two varieties i.e. the two forms of quatre-saisons violets that were extensively grown at the time. (see p.56, note 1)

56. quatre-saisons The word is explained by Millet. Originally applied to these violets as the only known kinds of single violets to be reliably remontant.

57. Parma violets never do Millet does not seem to have come across single forms of the Parma violet (but see 'Princesse de Sumonte'): nor does he mention the cv. 'Brandyanum', which is said to have been raised from seed, and appears to be a Parma violet. The earlier form of Parma violet (Neapolitan) was known to produce occasional single flowers late in the season. Single forms of Parma violet were said to be grown in Naples, and were listed in 1874 by Perkins of Leamington:- Neapolitan, single – 4/- a dozen.

58. Chapter 53 This is not by la Quintinye. See note 4.

59. violet in pyramid form M. violette en pyramide

60. storax This is the resin of *Styrax officinalis*, and sometimes of other related or similar trees, used as a fixative in perfumery. It was an important and familiar ingredient of many still-room recipes of the past. It appears in many of those given in E.S.Rohde's *The Scented Garden*.

61. more single than double flowers This is not what one would expect, as the seed is from cleistogamous flowers. From my own very limited experience seed of the hardy doubles 'Comte de Chambord' and 'Rose Double' produced plants with double flowers identical with those of their parents.

62. raw material M. à l'état réel

In *Historia generalis plantarum* (1586) it is said that violets were exported from Marseilles to Alexandria and other parts of Egypt for use in drinks and for medicinal purposes. The statement is in another section of viola, and not under sweet violet, but this is probably a mistake. It shows that trade of this kind had existed for a very long time.

63. Saint-Roch – a district of Nice; Villefranche-sur-Mer adjoins Nice; Grasse arrondissement – the district of Grasse, still the great centre for distillation of floral scents, including that of violets.

64. Second Empire (1852-1870) Napoleon I's nephew Louis Napoleon (1808-1873) seized power in a coup d'état. He was deposed on the 4th September 1870, two days after he suffered a humiliating defeat in his war against Prussia and was captured at Sedan.

65. Nice French territory for most of the 18th c. and under Napoleon, Nice was included in the Kingdom of Sardinia in 1815. In 1860 by the Treaty of Turin Savoy and Nice were ceded to France after a plebiscite. Garibaldi, who was born there, wanted to sail to Nice and hold it against the French, but was fortunately persuaded to set off for Sicily – and the rest of Italy – instead.

66. Alphonse Karr (1808-1890) A French writer whose best known gardening book is *A Tour Round My Garden* (English translation 1855). Letter XIV begins with a disquisition on the violet, and describes the multicoloured double violet 'Bruneau', which is probably no longer in existence. As Millet says, Karr vigorously promoted the flower trade in the Midi, made possible by railway transport to Paris and other markets. According to Phlipponneau the first consignment of flowers from the Côte d'Azur came to the Halles at Paris in November 1871.

67. Solignac (1846-1892) About 1870 Camille Solignac moved to Cannes for the sake of his health, and there introduced the use of glass for the development of the cultivation of a range of flowers on the slopes of hills near the Mediterranean coast. He played an important part in the growth of large-scale floristry there.

68. le Var This department is unusual in that the Var, after which it is named, no longer passes through it.

69. Wilson Edward Wilson (1813-1878) was born in London, and in 1842 emigrated to Australia, where he was successful in farming and journalism. He helped to found the *Australasian* in Melbourne in 1864 and returned to England the same year. He was a well-informed naturalist, established

a Zoological Society in Melbourne in 1861, and had the Queensland shrub *Eugenia wilsonii* named after him by a friend in 1865.

Some time between 1864 and 1868 he was in Oran (Algeria) and noticed a violet growing on the walls of the citadel. He was unable to get material for propagation before leaving, but had some sent to him in London. He then dispatched plants to his friend Ramel in Paris. Ramel, who had introduced some species of Australian eucalyptus to France, worked at the École de Médecine and he grew the violet in the gardens there.

By 1870 the violet, which Ramel had named 'Wilson' as a tribute to his friend, was being grown by Ramel, and also by Lavallée in his garden at Segrez. Lavallée later found what he thought was the same violet in Turkey. It was generally regarded as a form of *Viola suavis*.

Ramel made the violet available to growers in the Midi, where the climate suited it, and it was a great success commercially. Eventually it was superseded by 'Luxonne', a cross between 'Wilson' and 'Czar', and other new varieties. It is remarkable that it kept its name in the markets, so much so that for many years 'Wilson' was almost a synonym in the Midi for large-flowered violets.

There was much confusion about its origin. The name 'Violette de Constantinople' may come from inaccurate statements that it was found by Wilson (some said Ramel) not at Oran, but at Constantine (Algeria).

70. 'Czar' This cultivar was raised in England in 1863 by F.J.Graham of Cranford. Millet does not seem to have taken any interest in the development of "Russian" violets in England, which in some ways paralleled work on quatre-saisons varieties in France.

71. also called 'Reine Victoria' M. has 'Luxonne'

72. Mme.E.Arène Perhaps the wife of M.Arène, a customer of Millet's, who grew violets at Solliès-Pont in the Hyères district, and also exhibited fruit.

73. concours général One of the open shows of the SNHF.

74. Solliès-Pont M. Soliès-Pont

75. Haute-Garonne The department of which Toulouse is the prefecture.

76. Angoulême The mairie at Angoulême at present disclaims all knowledge of Parma violets. Millet listed 'Gloire d'Angoulême' as a synonym of 'Parme de Toulouse'.

158 *Armand Millet and His Violets*

77. Lalande M. Calande This is one of several misprints that suggest the printer was working from a manuscript copy which he had difficulty in reading, let alone understanding.

Violet kiosk at Toulouse at the beginning of this century.

78. Aucamville In 1961 they celebrated "le Centenaire de la Violette" at Toulouse. On that occasion it was said that the cultivation of Parma violets there began at St.Jory, extended to Lalande and Aucamville; then became established at Ginestous, Saint-Alban, Launaguet, Fontbeauzard, Fenouillet and Castelginest. At that time, 1961, there were 150 growers organised in two cooperatives.

Now this has all changed. The Parma violet at Toulouse has recently been restored to full health, is micropropagated and grown under glass in soilless cultivation. Under these scientific auspices and with well organised publicity the industry seems set to prosper.

One interesting suggestion is that the Parma violets grown at Toulouse may originally have been brought back from Italy by men who had fought there in the French army; just as some snowdrops came to Great Britain as a by-product of the Crimean War.

79. size of a five franc piece i.e. about 3.5 cms. (1⅜ ins.) in diameter.

80. wild radish M. ravenelle (that is *Raphanus raphinistrum*)

81. enfleurage M. effleurer for enfleurer – to carry out enfleurage. Enfleurage (the same word in French and in English) is: "The process by which 'Pomades' are made. A thin layer of lard or other fat is uniformly distributed on the surface of glass trays, which are then covered with fresh flowers. These are renewed periodically, until the fat becomes fully charged with the flower perfume. It constitutes the so-called 'pomade' from which the first, second, and third infusions or 'washings' are made". (Poucher)

82. Fresnes-les-Rungis Fresnes, with its prison that was notorious during the German occupation of 1940-1944, is immediately south of Bourg-la-Reine. Beyond it is Rungis, now the home of the flower market, having replaced the Halles. Millet's grandmother was born at Rungis and he had relatives there, so he must have heard a great deal about the cultivation of violets in the area. His father may even have worked in such violet fields when he was a boy.

83. éventaire There does not seem to be a word in English for this useful piece of equipment. Mayhew, describing street-sellers in London, called it "a tray slung round their shoulders". It is not surprising that violet-sellers were difficult to control in Paris. At Westminster Parliament rather ridiculously passed a law to prohibit the ringing of bells by muffin men, but to no avail. It was time, not politicians, that eventually silenced them.

84. First Empire Napoleon's rule as Emperor: 18th May 1804 to 6th April 1814. He was said to have fulfilled his promise to return with the violets when he left his place of

exile on Elba on the 26th February 1815 and entered Paris on the 20th March 1815.

85. in front of the espaliers   M. en contre-espaliers

86. other flowers being of little account   M. le reste des fleurs se confondait

87. Louis-Philippe (1775-1850) He reigned from 1830 to 1848. He was the son of the Duke of Orleans, called Égalité during the Revolution, who cast his vote for the execution of Louis XVI in 1792 and was himself guillotined in 1793.

88. Jean Chevillon Well known for his quatre-saisons violet, but nothing else appears to be remembered concerning him.

89. Vogt It seems most likely that it was to François Xavier Vogt, and perhaps to his father Joseph Vogt, that Alexandre Ruffin Millet owed his training in the specialist growing of violets and other crops with the use of frames.

François Xavier was born in 1811 at Ville St.Jacques (Seine et Marne), the son of a gardener, Joseph Vogt, who was then 38 years old, and Marie Geneviève Lansant. By 1825, when one of his children died, Joseph Vogt and his family were already living at Bourg-la-Reine.

In 1833 François Xavier married Marie Armandine Toron, and in 1836 a son was born who was named Xavier Harmand. In 1842 a daughter was called Armandine Madeleine. It seems probable that when in 1845 Vogt was a witness at the registration of birth of Alexandre Ruffin Millet's son, the names Harmand Joseph were chosen as a result of Millet's close relationship with the Vogt family.

Harmand appears to have been a usual spelling of the name in the first half of the 19th c. – at any rate in the registers of birth at Bourg-la-Reine – but like Armand Millet, Xavier Harmand Vogt later left off the initial H, as when witnessing his father's death in 1865 he gave his name as Armand Xavier Vogt.

90. Ravageot I do not know the origin of this name; perhaps it was called after the man who first found or grew it.

91. Sale was carried on wholesale in the rue aux Fers around the Marché des Innocents and the famous restaurants, fashionable at the time, Baratte and Bordier.   M. La vente s'en faisait en gros rue aux Fers autour du marché des Innocents, du fameux restaurant à la mode de ce temps, je nomme Baratte et Bordier:

rue aux Fers This is now rue Berger. It used to run beside the north wall of the Marché des Innocents.

Translator's notes

<u>Marché des Innocents</u> In 1786 the 12th c. cemetery of the Innocents was closed. The church was demolished and the whole cemetery dug over and the soil sieved. The bones were put in sacks and then transferred in covered wagons to the cemetery of Tombe-Issoire, and when this did not provide enough space, to other cemeteries in Paris.

The famous public fountain, which had stood against the wall and beside the church in the north-east corner of the cemetery, was carefully reerected in the centre, and the two blank sides ornamented with carvings to match as far as possible the original sculpture. It still served a useful function now that the cemetery had become a market.

The vegetable market that had been held in rue de la Ferronerie now took place a few paces further north in the Marché des Innocents.

There were about 400 large red parasols to provide shelter for the market women, and in 1813 wooden arcades were constructed against the surrounding walls to give additional protection. In the morning goods were sold wholesale, and then retail during the rest of the day.

On 7th September 1791, during the Revolution, the Constitution was proclaimed in the Marché des Innocents; in 1830 there was fierce fighting here during that revolution; and on 17th August 1850 an extravagant ball was held there, for which an enormous marquee was set up over and around the central fountain, and an elaborate entrance constructed.

It was along the north wall of the Marché des Innocents, in rue aux Fers, that flowers were sold until 1855, when this market was closed down and replaced by the nearby Halles. In 1969 the Halles ceased to function and the market moved to Rungis. Now only the Fountain of the Innocents remains in the middle of an unwalled open space.

<u>Baratte and Bordier</u> Baratte's restaurant was highly esteemed. Courtine said it was still open at 8, rue Berger in 1925. Bordier's, in a neighbouring street, was a restaurant for the market, popular and teeming with customers.

92. <u>without runners</u> M. <u>sans filets</u> This variety, 'Parme sans filets', does not seem to have found favour in England, but was always in Millet's list. As with forms of alpine strawberry that are equally restrained in their habits it made management easier. However, unlike the strawberries, it could not be reproduced from seed. Millet says it was known to growers in

The Marché des Innocents in 1855.

the Paris area from about 1840, but when shown in 1879 most people thought it was new. Difficulty of propagation may have caused it to be grown rarely except by specialists.

93. Paré I have not seen this name among births, marriages and deaths at Bourg-la-Reine during the nineteenth century. A nurseryman René Paré of Gentilly joined the Horticultural Society in 1855; he may have moved out to Bourg-la-Reine at some time.

94. Mascré, of Sceaux He was a general nurseryman, who won awards for his roses.

95. The Republic gave way to the Empire Napoleon III had been President of the Republic.

96. M.de Lavallée Alphonse Lavallée (1836-1884) became Secretary-General of the SCHF in 1873, its President in 1879. He was born in Paris. He farmed for a short time, but was passionately interested in botany, studying under Brongniart and Decaisne. In 1851 his father bought the Château de Segrez at Saint-Sulpice de Favères (about 40 km. SSW of Paris) and Lavallée began to plant what became a famous arboretum. In 1870 when the Germans arrived he sent his family away, but stayed to save the commune from pillage. In this he was successful. His early death came as a great shock to members of the SNHF.

97. one of the biggest growers of Verrières-le-Buisson. This refers to Semprez. In 1816 Philippe-André Lévêque de Vilmorin, head of the Vilmorin-Andrieux firm, bought an estate at Verrières-le-Buisson. I presume that this is why the violets were given to Semprez, a neighbour, for him to try out.

98. question... labour-saving M. une grosse question de main-d'oeuvre

99. Vilmorin The famous firm of Vilmorin-Andrieux has a very long and romantic history. It goes back to Philippe-Victoire Lévêque de Vilmorin who, in 1759 as a thirteen year old orphan, came to Paris to study. He became a friend of Pierre d'Andrieux, Botanist of Louis XV. Vilmorin was associated with the seed business of Andrieux and married his daughter Adélaïde. From then on members of the Vilmorin family directed this eminent French nursery.

100. 1870 It is not generally realised even in France what disastrous effects the German invasion and the siege of Paris had on the lives of nurserymen and market-gardeners in the Paris area.

101. Thomas Ware Thomas Softly Ware (c.1824-1901). From being a draper in Spitalfields he established a nursery at Tottenham in 1851. He introduced 'The Czar' into commerce.

102. Lemoine of Nancy Victor Lemoine (1823-1911). After many years working and studying in various nurseries he set up his own at Nancy in 1850. He was known familiarly as "le semeur" and raised a phenomenal number of widely acclaimed plants. A glance at the lilacs listed by Hillier's today shows the extent of his achievement. He gained many honours, including the Veitch medal. His firm was continued by his son, Émile Lemoine.

103. Yvon of Paris J.B.Yvon (1831-1903) made a renowned collection of herbaceous plants, and was famed for his displays at exhibitions in the years 1870-1900.

104. Marguerite de Savoie (1851-1926) The daughter of Ferdinand, Duke of Genoa, she married her cousin, who became King Humbert I of Italy in 1878. She seems the most likely claimant, an Italian riposte to the Austrian 'Marie-Louise'.

105. Marie-Louise Of the various contenders the most likely is Napoleon's second wife, Marie-Louise of Habsburg-Lorraine (1791- 1847). She was Duchess of Parma when, after Napoleon's death, she married Count von Neipperg, who was Grand Master of her palace at the time. The name would have found favour with Austrians, have pleased the French for quite different reasons, and seemed appropriate for a Parma violet in any case. The name 'Duchesse de Parme' which is commonly used for the usual form of Parma violet, in England at any rate, presumably also alludes to her. However, the name violette de Parme was already in use by at least 1805. At Udine they have another theory of the origin of the name.

106. Malmaison Rueil-Malmaison, the favourite residence of Napoleon when he was First Consul. Later Josephine made a famous garden there, renowned for its roses. English nurserymen were allowed to send plants during the war, being exempted from the blockade. The Château de Malmaison was built on the site of a leper-colony, and this accounts for its name.

107. 1821 It is mentioned in *Le Bon Jardinier* of 1805; Sweet's *Hortus Suburbanus* 1818; *Transactions of the Horticultural Society* 1820. There must be many references to the Parma or Neapolitan violet waiting to be brought to

notice in books, correspondence or diaries of the eighteenth century.

I have not yet seen an explanation of Collinson's 'Portugal Violet': "*Viola odorata* var. *Viola lusitanica flore caeruleo odoratissima*. The Portugal Violet, remarkable for its early flowering, sweet scent, and large yellowish green leaves, was, with the long-podded yellow wood Sorrel, raised out of mould sent from that country in a tub of plants by my ingenious friend Mr.Power. I have obliged many ladies with roots, who admire them for their fragrance; sent 4th May 1741." Collinson also noted: "it flowers about Christmas, and has a rich odour beyond ours". (*Hortus Collinsonianus* p.57)

108. one of the reasons   M. une des conséquences An involved sentence, but the meaning seems clear.

109. 'Parme de Toulouse': see note 78

110. Paillet This nursery was founded at Châtenay by Lelieur Paillet and continued by his son Louis (1837-1909), who was succeeded by his son Louis Paillet fils about 1892. Unfortunately he died prematurely in 1902 and the nursery was eventually taken over by Croux in 1911. They specialised in peonies and dahlias. In 1894 a commission from the Horticultural Society reported that their collection of herbaceous and tree peonies was "unique in the world".

111. 'Brune de Bourg-la-Reine'; 'Gloire de Bourg-la-Reine' In naming these violets Millet and his father publicised their home town, as did Margottin with his roses 'Belle de Bourg-la-Reine' and 'Gloire de Bourg-la-Reine'.

112. more deeply serrated   M. plus ramifié

113. 'Armandine Millet' Millet was particularly pleased with this cultivar that he named after his daughter.

114. fine rose   M. d'un beau rose à l'automne Perhaps like some blue primulas this variety gave flowers of a different colour in autumn, reddish tints rather than blue.

115. in a mosaic bed   M. en mosaïque Mosaical bedding was a term used for a time in England too, referring to patterned or carpet bedding. As 'Armandine Millet' had variegated leaves and was compact and neat, Millet thought it could be used in this kind of flower-bed, perhaps as an edging. *Le Jardin* (1896 p.184) gives designs for mosaiculture with a coloured illustration.

116. leaves variegated... blue flowers *Vinca minor* springs to mind. At first sight Millet's remark seems odd as there are

so many; but he means flowers suitable for edgings or mosaic beds. Few sweet violets with variegated leaves have been listed, but there have been some. Weston (*Universal Botanist* 1770) offered one with an unwieldy description and name: *Viola acaulis, foliis ex argenteo variegatis – odorata variegatifolia.* Several names from the past are given in Coombs (pp. 51-52).

117. Luchon; Luxonne These sound like place names, but the former can scarcely be Bagnères-de-Luchon. No doubt someone can explain.

118. One of my friends Claude Néant of Bièvres is the grower referred to but not named. He had raised the plant in 1868, but nearly lost it in the frantic period following the German invasion of September 1870. As Millet advised, Néant showed his rose-coloured Parma violet at a meeting of the SCHF on 28th October 1875. It was greatly admired, and given a 2nd Class award. One of the Committee, M.Verlot, said he failed to see how Néant could have obtained his plant from seed, because Parma violets never set seed. Néant offered to bring along a plant with seed capsules. On 11th November he showed a plant with its double flowers and capsules, as well as a bunch of the Parma violet flowers that became known later as 'Mme.Millet', thus conclusively vindicating his earlier statement. Millet acquired the new cultivar from Néant and in 1884 he marketed it under the name of 'Mme.Millet'.

119. 'l'Inépuisable' An intriguing parentage is given for this variety: Quatre-saisons X Russian ('Czar')

120. 'Swanley White'; 'Brazza White' Soon after 1878, when "Count Brazza's Neapolitan Violets" were illustrated in the *Gardener's Chronicle*, Cannell's nursery at Swanley bought two Parma violets from Comte Savorgnan di Brazza of Udine and named (or renamed) them. The white they called 'Swanley White', the purple ("reddish-purple with a mottled white centre") – 'Venice'. For some reason they soon abandoned 'Venice', but 'Swanley White' was a great success. Unfortunately for Cannell's it was generally known as 'Comte de Brazza', and some people insisted they were distinct varieties. In the end it was agreed that the white Parma violets grown under these two names were the same variety that Brazza had raised and Cannell acquired.

121. obtained from seed Count Brazza was reputed to have made deliberate crosses with Parma violets, but I do not

know whether he left any records about this. Millet's reference to the occasional appearance of violet-coloured flowers on 'Comte de Brazza' may support his assumption, but when he showed two cultivars as rare examples of sporting in violets, the other cultivar was 'Mme.Millet', which was definitely a seedling.

122. mutation M. anomalie Millet uses 'anomalie' also when he mentions the leaf colouring or variegation of 'Tigrée or'.

123. Forgeot An important firm of vegetable and flower seed merchants, with their shop in Paris at 8, quai de la Mégisserie. Their trial ground was at Vincennes. They frequently exhibited. Étienne Forgeot died in December 1906, aged 58.

124. influence of resin on the colouring of leaves It would be interesting to learn more about this phenomenon and the processes involved.

125. Épinay Either Épinay-sur-Orge (Essonne), which is south of Paris, and more likely as it is not far from Chilly-Mazarin, his father's birthplace; or Épinay-sur-Seine (Seine-St.-Denis) to the north of Paris, near St.Denis.

126. Maxime Cornu (1843-1901) One of the outstanding people in French horticulture and agriculture, known in England from the yellow tree peony that bears his name. In 1884 he became Professeur administrateur at the Muséum (Jardin des Plantes). He was a leading figure in the struggle against Phylloxera, when it threatened the whole French wine industry. He wrote extensively, introduced many plants and promoted botanic gardens in the French colonies.

127. Bruant of Poitiers F.R.Bruant (1818-1900) was the son of a gardener and established a large nursery centrally placed in Poitiers. He was succeeded by his son Georges Bruant, who further developed his father's interest in raising new varieties of a range of garden plants. After his death in 1912 the business continued under his son-in-law M.G. Viaud-Bruant.

128. 'Wellsiana' M. 'Welsiana' A violet raised by Mr.Wells, gardener at Fern Hills, near Windsor. In the Millet catalogue of 1932 it has become 'Velsiana'

129. 'Dybowski' Jean Dybowski (1856-1928) was the son of a Pole who came to France as a refugee after the Polish rebellion of 1830. Dybowski was educated at the École nationale d'agriculture. In 1889 he went to southern Algeria

on the first of his African explorations. He became Inspector General of colonial agriculture. In 1893 and 1894 he gave lectures at the SNHF on his expeditions to tropical Africa.

130. <u>Saint-Raphaël</u> – adjoins Fréjus

131. <u>Hyères</u>   M. <u>Hyre</u>

132. <u>Louis Achard</u> Louis Achard was referred to as "fleuriste à Hyères", as when in 1905, at almost the same time as Millet, he was given the award of Officier du mérite agricole for: "many awards at exhibitions... and thirty-six years of practical horticulture".

A quite different account of the origin of 'Princesse de Galles' from that given by Millet appears in *Les Violettes à Hyères* by Foussat. He says it occurred as a chance seedling among plants grown by Mme.veuve Recous, and that all plants of this cultivar at Hyères had been propagated from this one seedling; that the florist Achard had, with the consent of Mme.Recous, shown bunches of the violet at the meeting of the SNHF on 9th February 1893, as mentioned by Millet, and that it was named on that occasion. He also asserted that 'Gloire de Bourg-la-Reine' had never been grown on Mme.Recous's property. This complex matter can never be fully elucidated. Moreover, the report of the meeting on 9th February says that Achard stated that 'Princesse de Galles' had been found among his violets, and he made no mention of Mme.veuve Recous.

133. <u>'Gloire améliorée'</u> – that is to say an improved form of 'Gloire de Bourg-la-Reine'.

134. <u>'Amiral Avellan'</u> Named after the Russian Minister of Marine. In the period leading up to the Franco-Russian Agreement, on 13th October 1893, a Russian squadron under Admiral Avellan arrived at Toulon, and there were wild demonstrations of affection for Russia. The violet must have been named, or renamed, on this occasion.

135. <u>Léonard Lille</u> He was a distinguished horticultural seedsman who gave brilliant displays of a wide range of flowers at shows.

136. <u>Molin</u> Charles Molin died in August 1915, aged 64. He was best known as a seedsman.

137. <u>'Princesse Béatrice'</u> <u>'Comtesse Edmond du Tertre'</u> Millet came to have a higher opinion of these varieties later on, including them in lists of the most reliable and floriferous violets.

138. 'Princesse de Sumonte' *Gartenflora* (1895) has a coloured plate and an account by Ernanno Bredermeier, who found it "many years ago" in the garden of a villa in south Italy, where there were "six or seven plants" rather neglected. At the owner's request he named it after "a great lover of flowers, Principessa di Sumonti". Millet's catalogue has "genre Violette de Parme". I have left the French version of its name.

139. Cucullata violets M. cuculata
V. cucullata alba   = *V. alba*
V. cucullata grandiflora   = *V.obliqua*
V. cucullata striata   = *V. striata*
V. pubescens = *V. pubescens eriocarpa* (widely known as *V. pensylvanica*)

140. perennial tuberous violets – more correctly 'rhizomatous'.

141. Dugourd of Fontainebleau Pierre Dugourd (c.1831-1914) was a gardener ahead of his time. He devoted himself to hellebores and became known as "le père Hellébore". His plants were admired at the concours général each year, and even gained him a gold medal, but attracted little public interest. He began sowing seeds of his favourite plant in 1873 and continued raising new plants until his death. He was gardener to Count Circourt at Fontainebleau until 1888, and began his own nursery in 1889. Millet and Yvon were members of a commission that visited his nursery at his own request the same year. In 1896 he was awarded a silver medal for what must have been an interesting exhibit of hardy orchids.

142. Viola biflora At a casual glance the two flowers, and their leaves, are similar.

143. spreads out much more extensively M. est beaucoup plus développé dans son ensemble

144. more incurved M. plus en cornet

145. scent is of the sweetest For curious exceptions to this, Parma violets with a disagreeable smell, see *Revue Horticole* 1875 p.141 and 1884 p.102.

146. to their having the appearance of violets M. par l'esprit d'imitation

147. is free-flowering M. fleurit souvent

148. Consulate The French government from 9th November 1799 (the year 8 of the Republic) until 18th May

1804. At first Napoleon Bonaparte was one of three consuls nominated for a limited period; then he was made Consul for life. Finally in May 1804 the Consulate was replaced by the Empire.

149. in issues of Le Bon Jardinier we read M. Cependant de 1818 à 1820, les bons jardiniers nous parlent de la culture des violettes, des doubles, des simples et enfin des violettes de Parme, qui, disent-ils, doivent être mises sous châssis. This emendation may seem audacious, but elsewhere Millet tells us that Parma violets are first mentioned in 1821. Actually the 1805 issue of Le Bon Jardinier includes the Parma violet, but Millet cannot have had access to issues before 1818 except for that of 1793, which does not mention Parma violets. In 1893 M.Lecocq-Dumesnil presented the 1793 edition to the SNHF, and this handsomely bound book is in their library. It created a sensation; members were unacquainted with such an early issue and unusual format. This occasion must have impressed Millet.

In the 1805 issue Parma violets are said to be grown in pots in a glass-house (orangerie) or indoors. Frames are mentioned in 1806 and from then on.

150. Toulouse M. Toulon Toulon could be correct, to distinguish the Toulon and Hyères area from that of Nice and Grasse. However, Millet does not single out the Toulon region when dealing with Parma violets, whereas he devotes much space to Toulouse.

151. three colours Similar tastes prevailed in England at the time: "Ribbon borders... Public gardens generally confined themselves to three rows per ribbon – red, white and blue was a favourite pattern". (Elliott, Victorian Gardens p.130)

152. 'Rawson's White' The Rev.A.Rawson (c.1819-1891) bred florists' flowers, particularly pelargoniums. His white violet became a great commercial success. It was introduced in 1888 by Cannell's Swanley Nursery.

153. 'Viola à fleurs tigrées or' It came to be called simply 'Tigrée or' and remained in the Millet list, but does not seem to be known now.

154. fleuriste For foreigners the best known example of this use of the word is Le Jardin Fleuriste de la Ville de Paris, where plants are grown to supply flower-beds all over Paris. A well illustrated article about the new Fleuriste de la Ville de

Paris at that time appeared in the *Revue Horticole* in 1899 (pp.576-580)

155. strikingly effective M. d'un effet saisaissant (sic)

156. Chapter VII M. chapitre suivant avec les violettes de Parme

157. 'Neapolitan violet' M. violette de Naples Neapolitan violet was the name used in England throughout the 19th c. for the pale 'unimproved' form of Parma violet.

158. covered markets and street markets M. les halles et marchés There is a good account of the flower markets, and other ways in which flowers were sold in Paris – at florists' shops, at kiosks, from barrows – in Vilmorin's *Les Fleurs à Paris*. At that time, 1892, there were eleven flower markets in Paris, each opening on two or three days a week as well as important holidays. All wholesale trade was at the Halles. The earlier violet-sellers with their éventaires had become little more than a memory from the past.

159. for the rest of the year M. le reste de l'année i.e. until they lose their leaves in autumn.

160. the double violets 'Patrie' and 'en arbre' M. Les violettes doubles 'Patrie' et 'en arbre' At some time during the preparation of his book Millet decided that these were merely two names for the same violet, and regarded them as synonyms.

161. a strong green, but almost white where it joins the stem M. vert fort, blanc même, jusqu'aux pédoncules

162. so as not to scrape the rhizome or pull off roots M. afin de ne pas écorcher, ni égrener les racines

163. those that have turned yellow M. ainsi que les jeunes 'jeunes' must be a misprint for 'jaunes': compare his words on red spider, "elle fait jaunir les feuilles" (it causes the leaves to turn yellow) and (p.108) "a good few leaves turn yellow as a result of being transplanted".

164. of Parma violets, as well as that of single and hardy double violets M. des violettes de Parme tant simples que doubles Millet does not appear to have had any acquaintance with ordinary single Parma violets, although they were known in England and Italy at the time. (see note 57)

165. in a later chapter M. au chapitre suivant See Chapter XVII

166. outside the gardens M. elles sont toutes élevées en dehors des jardins. The word 'jardin' seems to be preferred to 'champ' when the area under cultivation was not extensive,

or when intensive work such as the use of frames or cloches was involved. Special propagation was carried out in a separate place.

167. galère It is a pity Millet did not give other examples of the language and customs of Parisian cultivators and market gardeners.

168. J.L.Desroches I have not traced Desroches, despite valiant efforts on my behalf by kind people in California and by other Ameriacan librarians. There is certainly no lack of large centres now; and the Ano Nuevo Nursery of Pescadero sells Parma violets at the San Francisco Flower Mart.

169. François Supiot In *The Florists' Exchange* (1894-1896) Supiot appears as a prominent grower of violets actively promoting the sale of the large single varieties from France – 'Luxonne', 'Amiral Avellan' and 'Princesse de Galles' – and winning awards for them at shows at Philadelphia and New York.

170. the ground M. en recouvrant nos terrains de cloches Perhaps 'terrains' is a misprint for 'terrines' – pots. The cloches were bell-shaped, sometimes called bell-glasses in English. They are illustrated and described in Macself's *French Intensive Gardening*.

171. if artificial fertilisation were easy to carry out M. si la fécondation artificielle eût été facile Darwin described cross-fertilising flowers of *V.canina* on 14th April 1863 and they had large capsules of seed by 3rd June, but I do not know whether he experimented in this way with *V.odorata*. There were one or two brief articles on the subject in gardening journals in the late 19th and early 20th c. Deliberate crosses were certainly being carried out by the time Millet's book appeared, but it seems unlikely that the nurserymen involved recorded their work systematically. Nor do we know when Millet began raising new varieties in this way.

172. À tout seigneur tout honneur This proverbial saying, which was a favourite of Millet's, goes back to at least the 17th c.

173. this rodent tribe M. cette gente rongeuse, instead of 'gent rongeuse'. 'Gent' is a feminine noun, used humorously after the manner of La Fontaine.

174. cyclamen In 1891 Millet was showing semi-double cyclamen, and reported that 60% of plants grown from their seed also produced semi-double flowers.

175. those who are not attacked... M. ceux qui n'ont pas à s'en défendre ont plus tôt fait

176. sow-thistle M. lasseron 'Lasseron' is a popular name for 'laiteron', the sow-thistle (*Sonchus oleraceus*).

177. healthy....flourishing M. saines et vermeilles cf. "le vermillon de la santé" (Dumas) Though Dumas was referring to complexion, this may explain the expression used by Millet.

178. Desmarets de Saint-Sorlin Desmarets (1596-1676) was born in Paris. He was one of the first members of the Académie, a dramatist, writer of verse, and habitué of the salons of the précieuses. This madrigal was commissioned for inclusion in the celebrated *Guirlande de Julie*, an album of floral paintings by Nicolas Robert accompanied by verses provided by leading writers of the day for each flower that was depicted. The *Guirlande* was presented to Julie d'Angennes in 1641 on behalf of her fiancé, M.de Montausier, before he left for the wars. The *Guirlande de Julie* was acquired by the Bibliothèque Nationale in 1989. A bunch of violets is portrayed on fol.34 and followed by two madrigals. The one by Desmarets is on fol.35; the second, on fol.36, is by M. de Mallerville:

> De tant de fleurs, par qui la France
> Peut les yeux & l'ame ravir,
> Vne seule ne devance
> Au juste soin de te servir:
> Que si la Rose en son partage
> Fait gloire de quelque avantage
> Que le Ciel daigne luy donner,
> Elle a tort d'en estre plus fiere,
> I'ay l'honneur d'estre la premiere
> Qui naisse pour te couronner.

It may seem incongruous to associate the "modest violet" with the language of the précieuses, though certainly not with the painting of Robert.

179. crystallised violets M. violettes candiées ('candi' seems to be more frequently used than 'candié')

These were made in the great houses in past times: "Apart from the home-candied rose-petals, violets, carnation-petals, cowslips... to be found in every well-regulated store-room in those days..." (*The Scented Garden*, by E.S.Rohde). Vernaut, of Boulevard Haussmann in Paris, was selling crystallised flowers in 1886.

180. Arnould This firm does not feature in the official

account of crystallised violets at Toulouse. This says that in 1897 a pâtissier, M.Viol, first made these sweets at his premises in rue Ogenne; that this enterprise expanded under his daughter and son-in-law, M.Bonnel. The present manufacturers, Candiflor, bought the clientèle in the 1960's. Bonnel, then at rue Perchepinte, exhibited at the Paris International Exhibition of 1900. No reference has yet been found to Arnould, but other makers of crystallised flowers who exhibited were Lapie (of Nice 1900), Olivier frères (of Toulouse 1904) and Rebours (of Troyes 1904).

181. Viola odorata sulphurea This violet was heralded as a novelty in the gardening press in France and in England. In October 1896 it was advertised in *The Garden* as on sale from L.Chenault, rue d'Olivet, Orléans, and was said to be the plant found by a postman of Indre. It had been recorded previously in France: *V.sulfurea* (Cariot, *Études des Fleurs*, ed.3, ii 63 1860); *V.odorata var. sulfurea* (Rouy et Foucaud, *Flore de France*, iii 26 1884). A similar violet named *V.vilmoriniana* was said to be distinct; and yet another form was found by an employee of Vilmorin's, A. Menassier, in the Trianon Park at Versailles. Although Vilmorin's expended time and effort trying to exploit the possibilities of this violet, they failed to achieve anything of note. Mrs.Gregory, in *British Violets*, mentions two similar violets found "in or near gardens" but "not planted". Other findings in the wild have since been recorded. It is usually distinguished from forms of *Viola odorata* by lack of scent, but the cultivar 'Irish Elegance' is said to be fragrant. It seeds about as prolifically as any violet, but does not seem to hybridise with other violets growing alongside.

When Millet showed it (24th February 1898) and the proceedings were printed in the Society's Journal, a note was given with a reference to Cariot's fifth edition, and saying that it had been found in the Loire and in Savoy.

In 1907 Mottet discussed its status and the failure of hybridisers to obtain interesting cultivars from the plants that had been found. (*Revue Horticole* 1907 p.514)

CULTIVARS

Cultivars raised or introduced by Armand Millet and his son Lionnel

Alba simplex
Armandine Millet (1878)
Brune de Bourg-la-Reine (1875)
Coeur d'Alsace (1916)
Explorateur Dybowski (1893)
Gloire de Bourg-la-Reine (1879)
Helvetia (1914)
La France (1891)
Lianne (1906)
Lilas (1876)
Marietta (1914)
Mlle.Bonnefoy
Mlle.Louise Tricheux
Mlle.Susanne Lemarquis
Mme.Laredo
Mme.Millet (1884) – seedling by Néant in 1868
Nana compacta (1903)
Opéra (1930)
Princesse de Galles (1889)
Rosea Delicatissima (1914)
Sans Pareille (1887)
Souvenir de Ma Fille (1912)
Souvenir de Millet père (1876) (raised by Millet's father)

As is seen with 'Princesse de Galles' the origin of some cultivars was disputed at the time. It is difficult to reconcile the two references by Millet in *Les Violettes* to the first appearance of 'Luxonne':
"Finally towards 1886-1887 there appeared in my seedbeds, almost spontaneously, a violet with large flowers, long petals, very long flower stems; in brief a magnificent variety that by chance came to be called 'Luxonne' ".
and "As it appeared at the same time in several places

where the two varieties were grown, no particular person can be said to have raised it. It was given its name by chance at the Central Markets".

Only later, when deliberate crosses were frequently being made, was there more certainty about the source of new violets, and even then little information was divulged.

VIOLETS LISTED

**Violets as listed by Lionnel Millet in his catalogue for
1932-1933
Millet, Domaine de Viroy, Amilly.**

Recent varieties from England:
Queen Mary
Lloyd George
John Radenbury

Recent varieties:
Coeur d'Alsace (Millet)
Mlle.Suzanne Lemarquis (Millet)
Mlle.Bonnefoy (Millet) (raised from Lilas)
Mme.Laredo (Millet) (raised from Amiral Avellan)
Opéra (Millet)
Président Poincaré (Parma violet)
Souvenir de ma Fille (Millet)

Amiral Avellan
Armandine Millet (Millet)
Askania (La France X B.de Rothschild)
Baronne de Rothschild
Brune de Bourg-la-Reine
California (raised from Luxonne)
Comtesse Ed.du Tertre
Cyclope (raised from Princesse de Galles)
Explorateur Dybowski (Millet)
Gloire de Bourg-la-Reine (Millet)
La France (type) (Millet)
La France
Le Czar (syn. Violette russe)
Le Czar Blanc
Lianne (Millet)
Luxonne
Mme.E.Arène (from Luxonne)

Mlle.Louise Tricheux (Millet) (raised from Czar Blanc)
Mlle.Schwartz
Mrs.Pinchurst (seems to be a cross between V.Odorata and Cuculata)
Noélie (originated in the Midi)
Princesse de Galles (Millet)
Princesse de Sumonte (Parma violet family)
Reine Victoria
Souvenir de J.Josse
Velsiana

Violets, second series.
Alba simplex (Millet) (from seed of the variety Blanche double)
Argentiflora
Le Lilas (Millet)
Mignonette
Mlle.A.Pagès
Nana compacta (Millet)
Odorata Rubra
Perle Rose (raised from the preceding variety)
Quatre-Saisons hâtive
Quatre-Saisons bleue
Quatre-Saisons Semprez
Rawson's White
Reine Augustine (from the mountains of the Tyrol)
Sans Éperon (curious rather than pretty)
Sulphurea
Subcarnea
Tigrée Or (in spring the leaves are variegated with white and bright yellow stripes)
Udine (it comes from the town of Udine and is similar to Reine Augustine)

third series – double violets
Belle de Châtenay (syn. Grandiflora tricolor)
Blanche double
Bleue double (syn. Princesse Irène)
King of Violet (syn. Louise Baron)
Patrie (syn. en arbre, La Parisienne)
Rose double (called de Bruneau)

4. – Parma violets
Comte de Brazza (syn. Swanley White)
Mme.Millet
Marie-Louise (syn. Marguerite de Savoie)
M.J.Astorg (raised from Marie-Louise)
Parme ordinaire
Parme de Toulouse (syn. Lady H.Campbell and Gloire d'Angoulême)
Parme sans filet
Président Poincaré

5
V. biflora
V. canadensis
V. cuculata
V. cuculata alba
V. cuculata grandiflora
V. cuculata striata
V. pubescens

The sixth and final section lists some varieties of what are called *V. cornuta* and *cornuta* de Paris. They were presumably garden Violas, one of them being Maggie Motte (sic). *V. cornuta* is one of the parents of the garden Viola.

TREE VIOLETS

There are three kinds of plant to which the name Tree Violet has been given at one time or another, and they should be considered separately. The first is one that grows upright of its own accord and was thought to resemble, or even to be of the same species as the sweet violet.

In classical Latin the word viola was used for several other plants as well as violets; stocks, wallflowers, and even some bulbous plants were included under this name. In the same way in England the word violet has been used to form popular names such as African violet, dame's violet etc. However, in the sixteenth century herbalists began to bring order into plant nomenclature and, aided by description and illustration they tried to restrict the use of viola to plants that seemed to belong to one family; and *Viola purpurea*, or *Viola martia*, was used for varieties of the sweet violet (now *Viola odorata*) and species similar in appearance (such as our wood or dog violets).

Matthiolus may have been the first to describe and illustrate a Tree Violet, in 1565 in his *Commentarii*. Here he named it *Viola arborescens*. In 1585 in *Discorsi* it appeared as *Viola arborea*, and in *Kreuterbuch*, published in 1719, at the end of a list of varieties of *Viola Martia* number 11 reads *Viola Martia arborea* ex Cyanaeo Albescente CB seu *Viola erecta* fl. albo *Hort. Eyst.*

Daleschamps in his *Historia generalis plantarum* (1586-7) said that this *Viola arborescens*, which was also called Mater violarum, had been sent to Matthiolus from Mount Baldo in north Italy, grew to a height of two cubits (about 40 inches), had many side shoots and had flowers like Consolida.[1]

Although it was clearly not closely related to the sweet violet, still less a form of it, it was sometimes listed under *Viola martia*, as for example:

---

1. Consolida 'Medieval name of a wound-healing herb, apparently from L. consolido, make firm.' (Smith and Stearn)

Viola martia arborescens
Viola martia arborescens lutea (Tabernaemontanus 1613)
and *Viola Martia arborescens purpurea: surrecta purpurea*
(The Upright Violet) (Sutherland 1683)

The name was now linked with a more useful descriptive one, which clearly showed that it was far from being a violet akin to *Viola odorata* : *Viola martia arborescens purpurea* (*V. elatior* Clusi) *Jacaea tricolor surrectis cauliculis,* quibusdam arborea dicta. (Commelinus 1689) i.e. The purple March violet growing in a tree-like form (the *Viola elatior* of Clusius) *Jacaea tricolor* with upright growing little stems, from which it is called the Tree Violet.

Fortunately Linnaeus eventually rescued readers from such lengthy descriptions and lists of synonyms by devising his binomial system of plant names.

The early association of Matthiolus' Tree Violet with the sweet violet lingered on long after *Viola arborescens* had been adopted as the botanical name of another species of Viola. Unsupported statements about sweet-scented violets growing wild in a tree or shrub like form were occasionally printed, as in the case of the 'Violette en pyramide' described by the anonymous author of *Instructions pour la culture des fleurs* that was included with later editions of la Quintinye's works. Millet continued believing in this "Tree Violet", even after Mottet, in the *Revue Horticole* in 1901, quoted the passage to show that it could not be a description of a violet.

In 1849, at the height of the popularity of trained Tree Violets, a lengthy and turgid article was published in *Paxton's Magazine:* 'On the adaptation of the tree violet for the early spring decoration of flower gardens, with a notice of its use for that purpose at his grace the Duke of Bedford's, Oakley.' By G.T. Although the author knew, and said, that the Tree Violet was so called because it was used for training in this form, and when left to itself grew in the same way as any other violet, he could not refrain from including at the beginning of his article: "It is supposed to have been brought a few years ago from China, where it is stated to assume a tree-like form three feet or more high." The entire article was at once translated into French and printed abroad, no doubt spreading this odd idea among its readers.

Finally, it is to be hoped, an unscrupulous nurseryman at

Bath, Edward Tiley, advertised in the *Gardener's Chronicle* (21st December 1850): "Tiley's Viola Arborea, or Perpetual Tree Violet... they bloom freely from August till the end of May, and are perfectly hardy. Twelve of these plants grown in pots will scent a large greenhouse... E.T. has been informed by the gentleman that introduced it into this country, that he has seen them growing in the thickets of Persia to the height of 4 feet, with large bushy heads to them, and hundreds of blooms on them at the same time. Large bushy plants, 6s. per dozen; smaller ditto, 3s. per dozen; or 1l. (i.e. one pound) per hundred."

During the eighteenth century orangeries, glass-houses and then conservatories had been developed and became widespread. They were used to enable exotic and tender plants to be grown and made it possible to introduce a whole range of exciting new species into European countries. Gardeners in botanic gardens and those working for well-to-do people learned how to manage them. Besides the more spectacular novelties there was a demand for pots of flowers in winter, especially fragrant flowers. One of the most important of these was the violet. With the violet the problem was not how to maintain a high temperature, but how to give them the mild shelter they needed in winter, maximum light and a good circulation of air at all times.

The Parma violet was included in *Le Bon Jardinier* in 1805. The description states that it flowers from October if kept under glass or in the house. In later issues frames are mentioned, and they came to be the usual answer to the problem, the pots being brought indoors when required.

One method of keeping violets healthy in winter when grown under glass was to induce the plant to form a short stem. By raising the crown from the soil it gave more air to the leaves and flowers, and apparently could lead to the formation of an increased number of flower buds.

Parma violets were ideal, but no other violets were then known that would bloom throughout the winter. It is not known when, but a double blue hardy violet appeared that could be expected to do so, and it was grown in pots in the same way as Parma violets. And as it was the only hardy double violet to be cultivated in the tree form at that time it came to be called 'en arbre' in France and the 'Tree Violet' in England. Millet gives it two other synonyms: 'Patrie' and 'La

Parisienne'. Curiously, it seems to have been the only hardy double violet ever to have had this ability to flower continuously from autumn to spring.

In an 1823 catalogue of F.Cels (Paris) we find: Viola arborea (orangerie), but without description. At about this time florists were developing the technique of training these Tree Violets, which reached their height of popularity between 1840 and 1860.

The results were not always entirely satisfactory aesthetically, to judge from the few illustrations we have, as the stem could be rather ungainly and large in proportion to the crown, and scars remained from where side growths had been rubbed out. One of the most attractive pictures of this kind of Tree Violet is that of 'Viola arborea Brandyana' (page 185)

One way to get the stem to grow to a reasonable height was to plunge the pot in soil and later remove the earth from around the new growth, taking off any roots that had appeared. Eight or nine inches seems to have been the usual height achieved rather than a foot or eighteen inches spoken of in the article by Beaton (appendix 1).

In April 1851 at the R.H.S. a Certificate of Merit was awarded to Messrs Hayes of Lower Edmonton for "two nicely-managed Tree Violets". Later on interest gradually declined.

When they had gone out of fashion people remembered them with affection, recalling their intoxicating perfume. In 1916 Mrs.Portman-Dalton wrote from Fillingham Castle about Tree Violets: "I remember them very well in greenhouses in Yorkshire fifty years ago. The stems were strong and woody, five to eight inches in height, never more. The Violets were beautiful and very sweet, a rich dark blue and very double, and with plenty of foliage and flowers. I have often regretted that they are never seen now. ....they lasted several years".

Colette was a writer unusually well informed about flowers. She wrote (of her mother): "Sido... used to lend books, receiving in exchange cuttings, and rooted plants of tree violets whose flowers were of a blue so dark as to be almost black, and whose bare trunk rose up out of the ground like that of a minute palm-tree".

Not everyone approved. In *Flora and Sylva* (1903) B.

(E.A.Bunyard?) – he may have been referring to Millet's methods, but his words are as applicable to those just dealt with – commented with acerbity: "Within the last few years there has arisen in some quarters a taste for Tree-Violets, in which infinite pains are wasted to induce an ugly and unnatural stem between the roots and the leaf crown: it is a pity that growers cannot find something better to do."

Armand Millet tried his hand at this form of Tree Violet early on in his career, but was dissatisfied with the results. It was with a different kind, using only new vigorous cultivars of single violets and training them with the use of a tall support, that he later succeeded.

On 11th November 1875, when Claude Néant confounded the Floricultural Committee of the SCHF by presenting a Parma violet plant bearing seed capsules, the members of the committee were not very gracious in admitting their mistake when they had said that Parma violets never bear seed. Instead one of them raised another subject. "M.A.Rivière[2] reported that double violets had formerly been sold at the markets as Tree Violets, the plants having a stem of 5 or 6 cms. As a child he had amused himself by growing violets in this form, removing their stolons and following the instructions given in gardening books. In this way, in the space of five or six years he had managed to get plants with stems of 15 – 20 cms, and he had kept plants for ten to twelve years. They flowered profusely and looked very pretty. He had achieved the same result with single violets."

Millet was perhaps present on this occasion, and as a member of the Society he would have read Rivière's words in the Journal. He knew Rivière, and as they were both interested in violets they almost certainly discussed the subject of training Tree Violets.

In 1906 in his note on Tree Violets Millet said they had first made attempts at selling them in pots about 1865, but without success; that in 1872 – 1875, using the new, vigorous varieties, their results were completely satisfactory; and that from 1896 they presented specimens 40-50 cms. high at the Agricultural Show.

---

2. He was director of the Établissement horticole de Bourg-la-Reine at the time.

The first attempts mentioned must refer to the old kind of plant, using double violets and having a fairly short stem. The reference to 1872-1875 is puzzling. He made no mention of the

A Tree Violet as trained in the mid-19th c.

practice in his book *Les Violettes*, written in 1895-1896 but not published until 1898. As stated above, the first date he gives for exhibiting a Tree Violet is 1896. When giving the ages of the violets he showed in 1906 the oldest is one of 'Mme.E.Arène' – 10 years – which again takes us back to 1896. This would seem to suggest that he began working with ideas similar to those voiced by Rivière in the 1870's, in common with some other growers, but encountered formidable obstacles. Only towards the end of the century, most likely, was he sufficiently satisfied with the results he had achieved to want to show plants and draw attention to them, explaining his methods.

In 1879 there was an article on Tree Violets in *Revue Horticole* written by May. This referred to the possibility of training a stolon of a violet vertically with the aid of a support, saying that in this way rapid results could be obtained. A further note later that year mentions heights of 40-60 cms.

In 1899 there was another article in *Revue Horticole* on Tree Violets, in which Dauthenay also emphasised the newer method of growing the plants to a considerable height with the help of a support.

Dauthenay reported that Georges Boucher, a nurseryman in the Avenue d'Italie in Paris, had shown Tree Violets between 15 and 50 cms. in height. However we do not hear of any other growers developing this line of cultivation to the extent that Millet and his son did.

In 1901 one of the regular contributors to *Revue Horticole*, M.S. Mottet, wrote yet another article on Tree Violets in which he said that for several years they had become a feature of horticultural exhibitions and had aroused much interest. However, he was concerned with the older, self-supporting kind of Tree Violet, which had apparently been forgotten by most people already. There was a picture of a gnarled specimen, 20 cms. high with a small crown and a few flowers. Although M.Boucher said he had once possessed the variety but since lost it, and others recognised it, no one appeared to be familiar with the way in which these earlier Tree Violets had been trained. And although Mottet quoted the account of the "Violette en pyramide" from *Instructions sur le jardinage* and said he did not think it could have been a violet, he still thought the description had been written by la Quintinye.

Tree Violets 187

It had been a long and arduous process to achieve the sizes Millet had in mind, and to develop the various shapes or forms he envisaged. But at last in 1900 he was able to show four plants to the Floricultural Committee and gained a 2nd

A violet trained in the table, or parasol form.

Class award for them. In 1902 he presented a plant of 'Mme.E.Arène' as a Tree Violet. It was four years old with a clear stem of 80 cms., the top growth 35 cms. in diameter and carrying more than 100 buds and flowers.

In 1906 Millet and his son Lionnel, who must be given credit for his share in the work of the nursery from 1900, showed 12 Tree Violets, 8 in the table (or parasol) form – the equivalent of a standard fuchsia or rose – and 4 as candelabras. He gave their ages, which ranged from one to ten years.

The next year they were awarded a gold medal at the Agricultural Show in Paris, held in March, for an array of violets, among them Tree Violets, set out around the orchid house.

From this time right up to 1914 Millet's Tree Violets were an annual attraction. He produced specimens one metre in height; he had forms that included a two-tiered parasol; and he gained a number of awards for them, one being a large silver medal in November 1910.

These extraordinary achievements, though not to everybody's taste, were undeniably a proof of exceptional skill in horticultural expertise and unremitting care. They were a flamboyant advertisement for his violets and his nursery; but in view of the enormous outlay in producing them, the constant attention they must have had, it is difficult to believe they were profitable.

In the spring of 1913, when displaying a fine array of Tree Violets, Millet and his son confidently asserted that it would be possible to have healthy plants twenty years old and more. On this occasion they gained a 1st Class award; and the following spring, when the award for their Tree Violets was reaffirmed, there can have been no suspicion of the terrible events in August 1914 that were to change the pattern of their lives for years to come.

The First World War seems to have practically put an end to the production of Tree Violets, just as the outbreak of the Second World War brought about the demise of the Millet nursery.

APPENDIX 1

Tree Violets. Every one is fond of violets, and if you had room for only three pots in the window, one of them should be a double violet. For seven or eight months of the year, or say from August to April, they should be in the window, and the tree violet is the best sort for pot culture. The French call it "The Perpetual Violet", which is perhaps the best name for it, inasmuch as that it flowers so freely and so much longer than any other violet. All violets may easily be trained so as to form little trees, as we call them, simply by bringing up a plant with one shoot only. This shoot should be tied to a neat stake, and all the side shoots be rubbed off as soon as they appear, unless you want to increase your stock of them: in that case, the side shoots may be left till they are three or four inches long, and then be taken off for cuttings. If these cuttings are planted at any time round the side of a pot in any light garden mould and watered, they will soon make roots. The best time, however, for increasing them by cuttings is the spring, and when they are well rooted they should be planted in the garden, and watered occasionally through the summer. They will make nice little patches, and begin to flower by the end of August; when a few of them may be taken up in succession and put into pots to bloom indoors all through the winter. If the tree violet is left to take its own way of growth, it will grow in patches, just like any other violet, without any attempt at forming itself into a little tree, and that is the easiest way to deal with it, and it is the way it produces the most flowers; but trained up in the tree fashion it looks very interesting, and will live many years. By the time it gets a clear stem a foot or eighteen inches high you may allow the side shoots above that heighth to grow, and then your miniature tree will be perfect.

From an article – *Window and Greenhouse-gardening* – by D.Beaton, *The Cottage Gardener* 1 (1848) p.48

APPENDIX 2

The Tree Violet

While several varieties of double Violets are generally esteemed and extensively cultivated, the real merits of the tree Violet are but little known. It is true that, under ordinary out-door cultivation it does not appear to possess attractions superior to other kinds; it even assumes a more prostrate form, and on this account it is often confounded with the old double blue Violet, from which it differs in several particulars, the principal being a perpetual habit of blooming, while its rival produces flowers at one season only. It is, therefore, as a pot plant, that the tree Violet becomes more especially worthy of attention; and under this kind of management, its profusion of flowers, and delightful fragrance, render it worthy of extensive cultivation.

The plan I have found eminently successful in its treatment is to take young rooted layers in April, and plant them in light rich soil, or a border having an eastern aspect. During the summer the plants are liberally supplied with water, and as they progress in growth all root-suckers and side shoots are removed. By the middle of September they may be taken up, potted into 5-inch pots, and placed in a cool frame, where in a short time they will commence blooming. As autumn advances I remove them to a light and airy part of the greenhouse, where they continue to flower until April; at that time they are shifted into pots a size larger than those they occupy and again receive the shelter of a frame. I prefer this season for the subsequent annual shifts. About the middle of May they are placed out of doors under a north wall, care being taken to prevent worms from getting into the pots, by placing them on a layer of coal ashes; all decayed foliage and suckers are removed, and if large plants are desired, it is requisite to take off all side shoots during this season. On the approach of autumn frosts the plants should be conveyed to their winter quarters, and treated as before. If due attention has been paid to keeping them in a healthy growing state, they will now be furnished with strong stems, 4 or 5 inches high, surmounted by a crown of large fragrant flowers; if necessary, the plants may be neatly staked, but under good cultivation supports will not be required.

When the season of potting again arrives, I shift into 8-inch pots, first carefully removing any unhealthy roots, or worn out stagnant soil; in the latter case it is preferable to shake away the whole of the ball, destroying as few fibres as possible; a tier of side shoots may now be allowed to proceed from the crown of the plant, these will naturally bend downwards to the edge of the pot; and a second tier being afterwards formed, as the crown advances in growth, fine pyramidal specimens from 12 to 15 inches in height will be obtained. When in perfection, these will be studded with flowers from the edge of the pot upwards. In subsequent shifts the ball should be carefully reduced, so as to allow repotting into the same sized pot as that the plant was growing in. I find 9 or 10-inch pots sufficient for the largest size; the plants may be annually shifted in these for some years with advantage.

The compost which I find the most suitable for this plant consists of two parts good turfy loam and one part well decomposed leaf soil, adding a sufficiency of sharp sand to render the material porous; during the more active season of growth, an occasional watering with clear manure water will be beneficial, and a sprinkling of clean water over head, during the heat of summer, will assist in keeping down red spider; should that pest, however, make its appearance, the pots should at once be laid on their sides, and the plants well syringed on the under parts of the leaves, as the ultimate beauty of the plant depends on the preservation of fine healthy foliage during summer.

Alpha.

(*The Gardener's Chronicle*, 28.2.1852, p.132)

APPENDIX 3

JSNHF **1902, P.144**

5. Non-competitive exhibit by M.Millet, nurseryman at Bourg-la-Reine (Seine): the Violet 'Mme.Arène' grown on a stem (Tree Violet)

In the note accompanying his presentation, M.Millet says: "This plant of 'Madame Arène' grown as a Tree Violet is four years old. It is in the table form, with a stem 80 centimetres high. The upper part is 35 centimetres in diameter and has more than a hundred buds and flowers. This plant has been shown as a non-competitive exhibit, the aim being to show what can be done with an ordinary Violet. The first years are the most difficult when trying to achieve this result, as the plant always has a tendency to send out growths from the base. When two and a half years old it passes into a shrub-like state and abandons, so to speak, its tendency to creep along the ground. To make it grow well it must be repotted twice a year, in October and in March. Its almost woody state makes the stem more delicate than when growing as a herbaceous plant. As soon as the first frosts appear it should be brought into a glass-house, where it must always be kept very close to the light.

Acting according to these principles you can obtain Violets of all forms, and very tall.

Single varieties are the most suitable, and those that succeed best of all are: 'La France', 'Comtesse Edmond Dutertre', 'Amiral Avellan', 'Madame Arène', 'L'Inépuisable', 'White Czar'.

(Vote of thanks)

APPENDIX 4

JSNHF **1906 p.104 8th February – 12 Tree Violets shown**
  p.205 <u>A note on Tree Violets</u> by Millet and son
    (Submitted 22nd February 1906)

With regard to the Tree Violets that we showed at the meeting of 8th February last, we think it will be useful to publish some information that is essential for preparing, forming and growing these plants.

At this meeting we showed ten vigorous examples of Violets grown in parasol form, and three in the candelabra shape.

Our fellow members of the Committee seemed to be interested in the specimens that we presented before them, and many thought that we were the inventors of this way of growing Violets as "trees".

This is not so at all. Indeed we find this method of cultivation described in an old book on horticulture: la Quintinye, 1730 edition, volume 2. There it is said[1] that if, at the beginning of winter, you pot up and bring into the orangery a kind of double Violet, 'Patrie', then under these conditions this Violet shows a tendency to produce a raised central stem. By training and raising this by means of a support you had the satisfaction in spring of seeing little rosettes of leaves appearing along the whole length of the stem that had been raised in this way; and afterwards the plants produced buds and flowers that filled the place where they were sheltered with fragrance. This note adds that after being grown in this way for three or four years, the plants might reach a height of 15 to 20 cms.

From this note you can see that nearly two hundred years ago the method of growing Violets in an erect form that we employ was already thought of; but then with the varieties that were available in those times it was, if not impossible, at least very difficult to manage to produce plants of large

---

1. For the actual words by the anonymous author in the original book see pages 43 and 44.

dimensions. This remained the case for a long time. It was only in about 1865 that we made some attempts at raising Violets on a stem and selling them in pots. We were not successful. It took too long to obtain a plant that was sufficiently developed. About 1872 – 1875, provided with vigorous new varieties with large basal stems, we made a fresh attempt, and this time we were completely successful.

From the year 1896 we exhibited several specimens of these Tree Violets, 40 to 50 cms. in height, at the Agricultural Show.

In 1900 we gave a large display, including several plants trained in the parasol form, and it is from these that the ones we showed recently in a state of vigorous health have been taken.

What has for a long time now been urging us on to continue our experiments is the difficulty there is in getting Violets to live and above all to flower in a glass-house, even one that is not heated, with normal methods of cultivation. When they are grown in pots in this way, the plants are too closely compacted and become weakly, producing only relatively small flowers.

When trained as a Tree Violet the plant puts out rosettes of leaves in tiers to right and left of the central stem, and this allows free access to air and light. Through this single stem, that has become woody, the sap is distributed more regularly, and encourages the development of flowers in succession throughout the whole winter.

Those who possess strong plants trained in the parasol or candelabra form can bring them into their rooms from time to time for a few hours; these rooms will then retain the fragrance for several days.

The cultivation of Tree Violets is quite simple. First of all vigorous young plants of the varieties that are most suitable must be obtained.

Among these we will mention: 'Amiral Avellan', 'Luxonne', 'Gloire de Bourg-la-Reine', 'Czar', 'Comtesse Edmond Dutertre', 'Princesse Béatrice', 'Baronne de Rothschild', 'White Czar', 'La France' etc.

If these plants are acquired in spring, they should be left to spend the summer in the open, the pots completely sunk in the ground. Then they are given supports to which the stems must be tied. The only other attention that is necessary is watering and keeping the plants clean.

About 15th September the plants will be brought into growth by cleaning up the surface of the pots: some soil will be taken off and replaced by good ordinary garden soil that does not contain too much manure. Care must be taken with watering, and any growths appearing at the base of the plant must be pinched out, as well as any that appear in unsuitable positions elsewhere.

Plants that you wish to train in the parasol form will be raised on a single stem, and pinched out at the height at which the top is to be formed. For the plants that you want to form into the candelabra shape you choose successively shoots well placed in tiers around the stem, taking care to keep the axis[2] as high as possible.

On 1st November the pots should be taken out of the ground and given a small layer of good, well-manured soil on the surface.

Throughout November the plants are to remain in their pots, out of the ground, and stay like that until the first frosts of -3° to -4° are experienced. This is the moment to bring them into the greenhouse, cool glass-house or under shelter.

During these various phases of cultivation the Violets that are to be trained into the parasol form will have developed at the crown growths in the form of stolons. This is the time to attach frames for training into the desired shape, to which the stolons will be tied. This work will have to be continued until the plant has been completely formed.

The diameter and height of the frame should be in proportion to the vigour of the plants.

An average diameter is 30 to 45 cms. A good average height is 50 to 75 cms. The attentions required for maintaining and growing on the plants are the same each year. Above all you must be careful not to leave roots out of the soil when renewing the earth in the pots. Treated in this way, Tree Violets will give complete satisfaction to those who produce them.

---

2 By 'axis' Millet means the central stem viewed in relation to the side growths, as a well-proportioned pyramid or candelabra was aimed at, but also one that was as tall as possible.

APPENDIX 5

JSNHF **1913 pp 98-99**.

At the Floricultural Committee:

1. Shown by MM.Millet, father and son, nurserymen at Bourg-la-Reine (Seine): 15 Violets grown in parasol form, from 60 cms. to 1 metre in height and 30 to 70 cms. in diameter, each one having from 150 to 200 flowers and buds; and 12 Violets grown in a branching, natural form, from 70 cms. to 1 metre in height, all furnished with flowers and buds. These plants are attractive, with their pleasantly irregular contours in many cases.

"When we first presented these Tree Violets," say M. Millet and his son, "some of our colleagues thought that it would be impossible to keep these plants alive for a long time, that they would die when their stems passed from the herbaceous state to a woody condition. This has not been so at all; plants between ten and twelve years old are the proof of this. What is more, we have noticed that the first and second year, when they are becoming adapted to these "tree" forms, they send out stolons around the base. In the third year, this tendency to put out runners has disappeared. The sap circulates in the woody part from the base to the summit, and a quantity of flowers develops, far exceeding what is observed on plants in a herbaceous state."

MM. Millet are convinced that with correct cultivation it would be possible to have Tree Violets twenty years old and more.

(First Class award)

APPENDIX 6

Bibliography of writings by Armand Millet, including brief notes

1874 *JSCHF* p.468 Ces Fraises, écrit M.Millet, ont été produites.....

1875 *JSCHF* p.145 (Roses forcées) Dans une lettre jointe à ces fleurs, M.Millet fils décrit de quelle manière il dirige et cultive les arbustes qui les ont produites.

1876 *JSCHF* pp. 726-727 Note sur les Haricots les plus avantageux pour la culture forcée.

1878 *JSCHF* pp.230-237 Note sur les différentes cultures de violettes aux environs de Paris

1879 *JSCHF* p.164 Fraisiers... plantés dans une caisse ou bac. M.Millet fait connaître le motif pour lequel il a essayé ce nouveau genre de disposition des Fraisiers à forcer.

1881 *JSNHF* p.182 M.Millet, horticulteur à Bourg-la-Reine (Seine), qui avait été chargé d'examiner un travail manuscrit de M. Quehen-Mallet sur la culture de la Pomme de terre...

1882 *JSNHF* pp.72-76 Compte rendu de l'Exposition de St.Maur-les-Fossées (13 août 1881)
JSNHF pp.313-318 Compte rendu de l'Exposition de Caen (13 octobre)

1885 *JSNHF* p.592 Compte rendu de l'Exposition de la Société d'Horticulture de Chartres

1886 *JSNHF* pp. L – LIII Congrès d'Horticulture de Paris en 1886: procès-verbaux des séances du Congrès. Mémoire de M.Millet de Bourg-la-Reine sur la 4e question (Influence de l'âge des graines sur les plantes qui en proviendront).

1887 *JSNHF* pp.153-154 Concombres obtenus dans une culture en serre.

1888 *JSNHF* p. 389 Note sur la Fraise 'Eléonore'
JSNHF pp.478-482 Compte rendu de l'Exposition Horticole ouverte à Nantes le 25 avril 1888
JSNHF p.632 Note sur deux pieds de Violette de Parme

qui portent des capsules.
1889 *JSNHF* p.56 Compte rendu sur l'Exposition d'Horticulture de la Dordogne, à Périgueux
Le Jardin p.42 Des Violettes
Le Jardin p.106 Culture des Violettes
1890 *JSNHF* p.451 Note sur sa Fraise 'Quatre-saisons améliorée'.
1891 *JSNHF* p.68 Note sur Cyclamens semi-doubles
JSNHF p.76 Note sur la Violette 'Mme.E.Arène'
JSNHF p.596 Note sur *Marenta juncea*
1892 *JSNHF* p.322 M.Millet ... dit qu'un membre... lui a rapporté voir... des Peupliers attaqués... par un gros ver...
1894 *JSNHF* p.77 Présentation de trois variétés de Violettes: 'Amiral Avellan', 'Princesse de Galles', 'Explorateur Dybowski'.
1896 *JSNHF* pp.123-124 La Violette 'La France'.
1897 *JSNHF* pp.136-139 Présentation de nouvelles variétés de Violettes, dont 'La France'...
JSNHF p.741 Note sur un hybride bigénéric... Helianthus X Harpalium
JSNHF pp.877-879 Rapport sur les cultures de Fraises de M.Emile Hennuy.
RH pp.472-473 Les Violettes et la variété 'La France' (planche couleur)
1898 *Les Fraisiers* (Doin, Paris: 218 p.; 52 gravures dans le texte)
Les Violettes, leurs origines, leurs cultures (O.Doin, Paris: 163 p.; 23 figures)
JSPF août et septembre 1898 pp.240-243 et 365-368 *Le Plébiscite sur les Fraises.*
1899 *JSNHF* p.51 Présentation de la Violette 'Princesse de Sumonte'.
JSNHF p.435-436 Allocution prononcée sur la tombe de J-B.Savoye.
JSNHF pp.1202-1206 Compte rendu sur l'Exposition de Colommiers.
1900 *RH* pp.122-123 Les Violettes sous verre (Reviewed in *JSNHF* 1900 p.303)
RH pp.665-666 Les Fraisiers remontants à gros fruits.
JSNHF pp.273-275 Rapport sur une visite à l'Etablissement de M. Rameau fils, horticulteur à Larue.

1902 *Le Jardin* pp.36-37; 88-90; 104-105. Les Fraises sur nos tables toute l'année.
JSNHF p.144 Violette 'Mme.Arène' cultivée sur tige.
1903 *JSNHF* pp.392-394 Rapport sur un ouvrage de M.Blanchouin intitulé *Le Fraisier : sa culture à air libre à la portée de tous.*
1904 *JSNHF* pp.611-612 revue de *La Culture de la Violette de Toulouse.*
1905 *RH* pp.44-45 Fraise du 'Quatre-saisons Millet' (planche couleur)
1906 *JSNHF* pp.205-207 Note sur les Violettes en arbre.
Le Jardin pp.88-90 Les Violettes de nos jours.
Le Petit Jardin pp.246-248 La violette dans les jardins.
1907 *JSNHF* pp.54-55 Rapport sur la deuxième édition d'une brochure de M.Blanchouin, ayant pour titre: *Les Fraisiers.*
JSNHF p.168 Rapport sur un opuscule de M.E.Jurgnet intitulé: *Le Fraisier*
RH pp.472-473 Variétés de Violettes (planche couleur)
1908 *JSNHF* p.110 12 Violettes en arbre de diverses formes en fleur depuis la fin d'octobre.
JSNHF pp.464-465 Quelques renseignements sur 20 variétés de Fraises remontantes à gros fruits et Quatre-saisons.
RH pp.416-417 Les meilleures variétés de Fraises remontantes (du *JSNHF*)
1909 *JSNHF* p.186 Note sur la Violette en arbre.
JSNHF p.515 Observations sur les Fraisiers remontants à gros fruits.
RH p.441 Les Fraisiers remontants à gros fruits (du *JSNHF*)
1910 *JSNHF* p.68 Rapport sur un opuscule de l'Abbé Touraine *Les Fraisiers Remontants*
JSNHF p.118 Note sur la Violette en arbre.
JSNHF pp.424-425 Fraisiers: 'Sybel', 'Londres 1908'; 66 vars.
RH pp.180-181 Dahlias parisiens (planche couleur)
RH p.491 Les fraisiers remontants à gros fruits.
1911 *JSNHF* pp.530-533 Compte rendu de l'Exposition internationale de Turin (17-24 septembre 1909)
RH pp.574-575 La Fraise Tardive 'Londres 1908' (planche couleur)

1912 *JSNHF* p.457 7 variétés de fraises.
RH. p.390 Fraisiers remontants à gros fruits (du *JSNHF*)
1913 *JSNHF* pp.98-99 Présentation de violettes en forme parasol et en faisceau.
JSNHF pp.115-119 Les Fraisiers remontants à gros fruits en 1911 et 1912.
JSNHF p.382 Rapport sur les additions à la deuxième édition de l'ouvrage de M.Blanchouin, ayant pour titre *Les Fraisiers*.
RH pp.158-161 Les Fraisiers remontants à gros fruits en 1911 et 1912 (du *JSNHF*)
JSPF pp.263-270 Remarques sur les fraisiers.
1914 *JSNHF* p.91 Nouvelles variétés de Violettes.
RH p.135 Nouvelles variétés de Violettes de MM.Millet et fils.
1915 *RH* pp.492-493 Les Fraisiers remontants en 1914.
1916 *JSPF* pp.73-76 Observations sur les Fraisiers et sur la production des Fraises en général pour 1915.
1917 *JSPF* pp.155-157 Les Fraises en 1916.
1920 *JSNHF* P.79 Note sur la Violette 'Coeur d'Alsace'

Obituaries

1920 *JSNHF* p.258
RH p.155
Le Jardin p.131
La Rive Gauche 21-8-1920

1921 *JSPF* pp.43-44 (Nomblot)

BIBLIOGRAPHY

For Dodoens:
Anderson, F.J. *An Illustrated History of Herbals* (N.Y. and Columbia University Press 1977)
Arber, A. *Herbals, Their Origin and Evolution* (C.U.P. 1938)
Bailey, K.C. *The Elder Pliny's Chapters on Chemical Subjects* (Arnold Part 1 1929, Part 2 1932)
Costaeus, J. *Commen. ad Messuen* (Venice 1581)
Dodoens, R. *Florum et coronariarum odoratarumque nonnullarum herbarum historia* (2nd edition 1569)
Dodoens, R. *Stirpium historiae pemptades sex* (1583)
Loeb Classical editions of **Athenaeus, Pliny, Theophrastus, Virgil, Vitruvius** etc.
New Century Classical Handbook ed. C.B.Avery (Harrap 1962)
Oxford Classical Dictionary ed. N.G.L.Hammond & H.H.Scullard (O.U.P. 1970)
von Haller, A. Bibliotheca Botanica (1772)

For la Quintinye:
la Quintinye, J.de *Le Parfait Jardinier ou Instructions...* (Paris 1695)
la Quintinye, J.de *Instructions pour les Jardins Fruitiers et Potagers* (nouvelle édition 1730 Paris)
Warner, M.F. and **Brown, J.S.** *Early Horticultural Literature* (Washington 1939)

For Bourg-la-Reine and the Siege of Paris:
– *État des communes à la fin du XIX siècle: Bourg-la-Reine* (1899)
Horne, A. *The Fall of Paris* (1965)
Lieutier *Bourg-la-Reine, essai d'histoire locale* (1914)
Roth, F. *La Guerre de 70* (1990)

For the Cultivation and History of Violets in France:
Carré, A. *La Violette de Toulouse* (Montpellier 1909)

Foussat, J. *Les Violettes à Hyères et dans la Région Méditerranéenne* (Hyères 1908)
Lagarde, J. *Culture de la Violette de Toulouse* (Toulouse c.1904)
Le Bon Jardinier 1805
Phlipponneau, M. *La Vie Rurale de la Banlieue Parisienne* (Paris 1956)
Ragonot-Godefroy *La Pensée, la Violette, l'Auricule...* (Paris 1844)
Vilmorin, P. *Les Fleurs à Paris – Culture et Commerce* (Paris 1892)

Other Subjects:
Coombs, R.E. *Violets* (Croom Helm 1981) (also many articles in *The Plantsman* and other journals on violet-growing in Great Britain)
Courtine, R. *Le Ventre de Paris* (Paris c.1928)
Darwin, C. *Different Forms of Flowers on Plants of the Same Species* (1880)
Desmond, R. *Dictionary of British and Irish Botanists and Horticulturists* (Taylor and Francis 1977)
Elliott, Brent *Victorian Gardens* (Batsford 1986)
Karr, A. *A Tour Round My Garden* (Routledge 1855)
Macself, A.J. *French Intensive Gardening* (Collingridge 1932)
Masters, M.T. *Vegetable Teratology* (Ray Society London 1869)
Percival, M.S. *Floral Biology* (Pergamon 1965)
Poucher, W.A. *Perfumes, Cosmetics and Soaps* (Chapman and Hall 8th edn.1974)
Rohde, E.S. *The Scented Garden* (1931)
Smith, A.W. and **Stearn, W.T.** *A Gardener's Dictionary of Plant Names* (Cassell 1971)

Horticultural Journals, especially:
JSCHF / JSNHF, Le Jardin, La Revue Horticole (RH), The Gardener's Chronicle, The Garden

ACKNOWLEDGEMENTS AND THANKS TO

Dr.Brent Elliott and the other librarians at the Lindley Library, which has been my main source of information.

Alain Renouf, Ervan Boudard and Latifa Mouaoued at the Library of the French National Horticultural Society.

Philippe Chaplain (Archivist at Bourg-la-Reine).

Madame de Girodon (Librarian at Bourg-la-Reine).

The staff at the Departmental Archives at Nanterre (Hauts-de-Seine), Corbeil-Essonnes (Essonne) and Créteil (Val-du-Marne).

Madame Madore and Madame Virole (Bibliothèque historique de la Ville de Paris), who have been most generous with their time.

Madame Gousset (Bibliothèque Nationale).

Madame de Nave (Plantin Museum, Antwerp)

Madame van de Ponseele (Library of the Muséum, Paris)

Monsieur Joseph Maréchal, for his kindness in responding to my appeal in *Jardins de France.*

Madame Girault of Orleans for her interest in the work.

Monsieur and Madame Croux, Monsieur Nomblot and Monsieur Traversat for replying to my enquiries.

Fiona Piddock (Librarian, Lincoln College, Oxford) for prompt assistance with classical texts.

R.I.Ireland (Department of Greek and Latin, University College, London) for painstakingly explaining phrases in Dodoens.

Dr.Robert Tombs (St.John's College, Cambridge) for advice on reading material about the Siege of Paris and responding to historical queries.

Janet Evans (Pennsylvania Horticultural Society) for information about François Supiot; Anita Karg (Hunt Institute); Barbara Pitschel (Strybing Arboretum) and other librarians in the U.S.A. who were helpful.

Mia Amato for information about the Flower Mart at San Francisco.

Enez Lesser and the late Dr.James Willis, of Melbourne for details about Edward Wilson.

Dr.Alan Leslie for his patience and good humour.

Stephen Taffler for drawing on his fund of knowledge about variegated plants.

Andrée and René Verheylewegen for assistance with the text, and details of the liner Aquitaine.

Julian, who untiringly sought information on my behalf, and Patricia, who unfailingly supported me at all times.

INDEX

Achard, Louis 15, 26, 59, 168
Actuarius 39, 152
Agricultural Show 184, 188, 194
Alexandria 155
Alpha 191
Amilly 23
Anderson, F.J. 151, 152
Andrieux, Adelaïde d' 163
    Pierre d' 163
Angennes, Julie d' 173
Angoulême 49, 157
Ano Nuevo Nursery 172
Antony 3
Aquitaine, the liner 17
Arabic, Arabs 152
Arène, M. 157
Aristotle 148
Arnould 140, 173
Aster amellus 153
Aster atticus 40, 153
Athenaeus 149
Attic ochre 37, 38, 150
Aucamville 49, 158
Australasian, The 156
Avellan, Admiral 168
Avicenna 152

Bagneux 3
Bagnolet 51
Baloche, Marie Germaine 1
    Pierre Ruffin 1
Baratte, restaurant 54, 160, 161
beans, French 9, 10, 135
Beaton, D. 183, 189
Belgium 7
Bertelot, Marie Rosalie 7, 11
Bibliothèque Nationale 147, 173
Bièvres 166
Bois de Boulogne, de Clamart etc. 50
Bon Jardinier, Le 82, 154, 164, 170, 182
Bonnel, M. 174

Bordier, restaurant 54, 160, 161
Borgue, J.B.F. 3, 7
Boucher, Georges 186
Bourg-Égalité 1
Bourg-la-Reine,
    conseil municipal 8
    enemy occupation 8
    Grande Rue 7, 25
    Horticultural Shows 1902 and 1905 17
    Petit Bagneux 25
    presbytery 24
    railway 3, 17, 25
    rue Alfred Nomblot, rue Armand Millet etc. 24
    voie du Port Galand 25
Brazza, Count 154, 166
Bredemeier, Ernanno 169
Bréviande, F.N.D. 2
Brioude 23
British Violets 174
Brongniart 163
Bruant 67, 167
Bruneau 2, 7, 23
bunches, montés plats 70
Bunyard, E.A. 184
Burelle 27, 28

Candiflor 174
Cannell, H. 166, 170
Cannes 156
Cariot 174
Castelginest 158
Cels, F. 183
Centenaire de la Violette, le 158
Charonne 51
Châtenay 7, 53, 55
Chenault, L. 174
Chevillon, Jean 53, 160
Chevilly 3
Chevreuse 112, 137
Chilly; Chilly-Mazarin 1, 8
Circourt, Count 169

205

cleistogamy 146
Cobbett, W. 155
Colette 183
Collinson 165
Commelinus 181
Commentarii 180
consolida 180
Constantine 157
Constitution 161
Consul, First 164
Consulate 81, 169, 170
Cornu, Maxime 67, 167
Costaeus 152
Cottage Gardener, The 189
Courtine 161
Cratevas 153
Croux, nursery 7, 19, 165
Cruydebook 151
crystallised violets 140, 141, 173
cucumbers 9, 10, 12
cyclamen 14, 123, 131, 172

dahlias 17, 18
Daleschamps 180
Darwin, Charles 172
Dauthenay 186
De Architectura 37
Decaisne 163
Delabergerie, D. 2, 23
De Materia Medica 153
Desmarets de Saint-Sorlin 139, 173
Desroches, J.L. 120, 172
Dioscorides 40, 153
Discorsi 180
distillation 49
Dodoens, Rembert (Rembertus Dodonaeus) 38, 40, 44, 147, 148, 150, 151, 153
Doin, Octave 15, 33, 34, 147
doubling of violets 41, 42
Doué 8
drosaton 39, 152
drying violets 51
Dugourd, P. 72, 169
Durand, Didier 2, 3, 23
Dybowski, Jean 67, 69, 167

École de Médecine 157
École Nationale d'Agriculture 167
Elba 148, 160
Empire, First 52, 54, 59, 82, 159
Empire, Second 46, 55, 116, 156, 163

enfleurage 51, 159
Enquiry into Plants 148
Épinay 66, 167
Eretrian earth 38, 150
essential oils 47, 49
Études des Fleurs 174
Eugenia wilsonii 157
Evelyn, John 147
éventaire 52, 159, 171
Exhibitions:
  Bourg-la-Reine 17
  Brussels 17, 19
  Buffalo 17
  Caen 12
  Chartres 12
  Chelsea (International Horticultural) 20
  Franco-British (London) 18, 19
  Ghent 17
  Milan 17
  Nancy 18
  Nantes 12
  Paris: 1877 10
    1878 10
    1879 11
    1880 11, 12
    1889 14
    1900 28
  Périgueux 12
  Roubaix 17
  Saint Louis (Michigan) 17
  St.Maur-les-Fossées 12
  Saragossa 17
  Turin 17.

Fenouillet 158
fleuriste, jardin fleuriste 25, 88, 170
Fleurs à Paris, Les 171
Fleurs au XIXe Siècle, Les 146
Flora and Sylva 183
Flore de France 174
Florists' Exchange 172
Florum et coronariarum... 147, 151
Fontainebleau 169
Fontbeauzard 158
Fontenay-aux-Roses 3, 53-55, 64, 82, 115, 117
Forêt de Bondy 50
Forgeot, Étienne 66, 73, 76, 167
Fountain of the Innocents 161
Foussat 168
Fraisiers, Les 147, 148

Franco-Russian Agreement 168
French Intensive Gardening 172
Fresnes 159; Fresnes-les-Rungis 51-53, 55, 81, 159

Galen 39, 152
Garden, The 174
Gardener's Chronicle, The 19, 166, 181, 191
Garibaldi 156
Gartenflora 169
Gaudichau, Jeanne Henriette 15, 16
    Joseph Henri 15, 16
    Marguerite 16
Gentilly 163
geoponici, georgici 149
Georgics 149, 153
Gerard, John 149, 152
Gerard's *Herbal* 147
Gesner, Conrad 151
Ginestous 158
gladioli 13, 25
Graham, F.J. 157
Grasse 45, 46, 156, 170
Guirlande de Julie 173

Haller, A. von 149, 152
Halles 54, 156, 159, 161, 171
Haute-Garonne 49, 157
Heim, Dr. 15
hellebore 169
Hermolaüs Barbarus 36, 149
Historia generalis plantarum 155, 180
Historia naturalis 148, 149
Holland, Philemon 150
Horticultural Congress 13
Hyères 15, 68, 69, 157, 168, 170

Illustrated History of Herbals 151
Indre 149, 174
Instructions pour la culture des fleurs 181
Instructions pour les jardins fruitiers et potagers 147
Instructions sur le jardinage 186
Io 36
ion 36, 149
Ion 149
iris 13, 15, 20, 23

Jacaea tricolor 181
Jamin, Ferdinand 23

Jean Laurent 2, 3, 8, 23
Jardin, Le 14
Jost 2, 23
Joubert de l'Hiberderie Prize 15, 146
Joubert de l'Hiberderie, Dr. 146
Juliana Anicia Codex 153

Karr, Alphonse 46, 47, 156
Kreuterbuch 180

Lalande 49, 158
Lapie 174
Lapierre, F. 9
la Quintinye, Jean de 33, 40-42, 44, 147, 148, 150, 154, 155, 181, 186, 193
Larue 3
Larue-Chevilly 7
Launaguet 158
Lavallée 56, 157, 163
Lavisé 1
L'Haÿ 3
Lecocq-Dumesnil 170
Légion d'honneur, Chevalier de la 19
Lemoine, Victor 7, 59, 164
Leo Africanus 152
Lille, Léonard 70, 168
lily of the valley 14
Linnaeus 181
Lobel 153
Loire 174
London, George 147
Longjumeau 8
Louis
    XIII 44, 80
    XIV 44
    XV 163
    XVI 160
Louis-Philippe 53, 54, 160

Macself, A.J. 172
Mallerville, de 173
Malmaison 59, 164
Marché des Innocents 54, 160, 161
Marcilly, Clarisse Joséphine 3, 13
Marcoussis 116, 117
Margottin, J.J. 2, 12, 23, 165
Margottin fils (Jules) 11, 23
Marguerite de Savoie 164
Marie-Louise 164

Martyn 154
Mascré 55, 163
Massy-Palaiseau 51
Masters, M.T. 155
Mater violarum 180
Matthiolus 180, 181
May 186
Mayhew, H. 159
Mechlin (Malines) 151
Melbourne 156
melons 9, 10, 13, 17
mérite agricole, Chevalier du 16
	Officier du 17, 168
Messues 39, 152
Millet, Alexandre Ruffin
		(Millet père) 1-3, 8, 12, 14, 160
	Armand Joseph (Harmand
		Joseph) passim
	Armandine Rosalie 7, 15, 16, 19, 28
	Auguste Lionel 8, 12
	Clarisse Joséphine (see Marcilly)
	Jacques Saturnin 1
	Jean Baptiste 1
	Lionnel Edmond 12, 15, 16, 20, 23, 177, 188
	Marie Rosalie (Varengue) 10, 17
Molin, Charles 71, 168
Monprofit O. 17
Montausier, M.de 173
Montfermeil 18
Montreuil 52
Montrouge 9
mosaic flowerbed, mosaiculture 62, 66, 165, 166
Mottet, S. 154, 174, 181, 186
Mount Baldo 180
Muséum de Paris 67
Mycon 150

Nancy 164
Naples 154
Napoleon I 1, 35, 148, 156, 159, 164, 170
Napoleon, Louis (Napoleon III) 156, 163
National Guard 8
Néant, Claude 27, 63, 166, 184
Neapolitan violet 154, 164, 171
Neipperg, Count von 164
Netherlands 7
Neuilly 16

New York 17, 120, 172
Nicander 36, 149
Nice 46, 47, 68, 140, 156, 170
Nicolai, Arnold 161
Nicolas 146
Nomblot, Alfred 2, 17, 23

obituaries, list of 185
Olivet 2
Olivier frères 174
omotribo 39, 152
omphacinum 152
Oran 157

Paillet 60, 62, 165
Palaiseau 115, 116
Paré 54, 163
Paris, cemetery of the Innocents 161
	Champ de Mars 10
	quai de la Mégisserie 167
	rue aux Fers 54
	rue Berger 160
	rue de la Ferronerie 161
	rue de Rennes 8
	rue du Four 8
	St.Germain-des-Prés 8
	St.Sulpice 8
	siege of 8, 80, 163
	Tuileries 9
Parma 91, 154
Parma, Duchess of 164
Paxton's Magazine 181
peonies 13, 18, 25
perfume 45, 47, 49, 56, 82
Pergamum 152
periwinkle 50, 58
Perkins 155
Petit Jardin, Le 147
Philadelphia 120, 121, 172
Phillips 154
Phliponneau 156
phlox 13, 25
phylloxera 167
Plantin 147
Plebiscite on Strawberries 16
Pliny 36, 40, 148-150, 153
Polygnotus 150
Portman-Dalton, Mrs. 183
Portugal 154
Portugal violet 165
potato cultivation, ms. on 12

Poucher, W.A. 159

Queen Victoria 15, 61

Ragonot-Godefroy 155
Ramel 157
Ravageot 160
Rawson, Rev.A. 170
Rebours 174
Recous, Mme.veuve 168
resin 66, 67, 167
Revue Horticole 13, 17, 154, 169, 174, 181, 186
Rivière, A. 23, 184
Robert, Nicolas 173
Robine, Athanase 9
Rochford, nursery 18
Rohde, E.S. 155, 173
Romainville 81
roses 9, 35, 46, 47, 53, 133, 165
Rouy et Foucaud 174
Rueau, M. 17
Rungis 1, 159, 161
Russian violets 157

Saint-Alban 158
Saint-Cloud 51
St.Jory 158
Saint-Raphael 68, 168
Saint-Roch 45, 156
St.Vincent de Paul, Sisters of 8
San Francisco 120, 121, 172
Savoy 28, 156, 174
Savoye, J.B. 16
Sceaux 3, 9, 55, 82, 115, 117, 163
Scented Garden, The 155, 173
scorpions 153
Sedan 8, 156
Segrez, Château de 56, 157, 163
Semont, Château de 147
Semprez 57, 163
serapion 39, 152
Serapion 152
Servius 36, 149
Severus 152
Shepherd's Bush 18
sil 150
Solignac, Camille 47, 156
Solliès-Pont 48, 157
sow-thistle 133, 173
stock 44, 47
storax 43, 155

Stratford-upon-Avon 154
strawberries 9-11, 13-16, 19, 20
stucco painters 150
Sumonti, Principessa di 160
Supiot, François 120, 172
Sutherland 181
Sweet's *Hortus Suburbanus* 164

Tabernaemontanus 181
Theophrastus 36, 148
Thoméry 17
Tiley, Edward 182
tisane 36
Toulon 47, 168, 170
Toulouse 49, 83, 140, 158, 159, 170, 173
Tour Round My Garden, A 156
Transactions of the Horticultural Society 164
Tree Violets 16, 27, 40, 41, 180-196
Turin, Treaty of 156
Turkey 56, 91, 119, 157
Turner, William 151

Udine 154, 164
United States 17
Universal Botanist, The 166

vaccinium 36, 37
van der Borcht, Peeter 147
van Kampen, Gerard Janssen 147
Var le 47, 156
Varengue, Antoine 7
    Marie Rosalie 7, 15
    Robert Jean Cyr 7, 12
variegated leaves 66, 67
Vegetable Teratology 155
Veillées Horticoles 146
Veitch, H. 18
    nursery 18, 19
    medal 164
Verlot, M. 166
Vernaut 173
Verrières-le-Buisson 55, 56, 115, 117, 163
Versailles, Royal Gardens at 40, 174
Victorian Gardens 170
Villard 146
Ville St.Jacques 160
Villefranche-sur-Mer 45, 156
Villejuif 7
Vilmorin 16, 57, 163

Vilmorin, Philippe-André 163
  Philippe-Victoire 163
Vincennes 51, 167
vines 1, 136
Viol, M. 173

Violets, Cultivation
areas: Angoulême 50;
  Central and Northern France
    and abroad 112
  Marcoussis 116
  the Midi 45, 46, 115, 118, 119
  Paris 51-53, 112-115
  Spain, Italy and Turkey 119
  U.S.A. 120, 121
beds 81, 86, 87, 96, 97, 105
borders 84-86, 88, 89
bunches 57, 58, 70, 117
cloches 126
cucullata violets 72, 73, 75, 76, 94, 95, 103, 104
diseases and treatment 136-139
division 81, 86, 97, 99, 103, 116
double violets, hardy 100-102
  at Chevreuse 112
forcing 50, 106-114
forcing on the spot 110, 111
frames: at Angoulême 50
  at Toulouse 49, 119
  in the Midi 119
  dimensions 106
  use of old frames 101
  single varieties for use in 89, 90
greenhouse 123
heating from the paths 50, 107, 110, 113
mats 87, 88, 106, 107, 113
mulching 101, 117
pests and countermeasures 130-135
planting 84, 97
pollination, artificial 127, 128
pots, violets in 93, 94, 100-102, 110, 122-124
prices 45-47, 55, 57, 58, 96, 114, 115, 117
propagation (see division, runners, seed); of 'Parme sans filets', 'Patrie', and 'Blanche de Chevreuse' 102
renewal of plants 88
runners, removal and use of 98-102

seed: harvesting 125
  sowing 82, 95, 126
  stratifying 126
transplanting 109
ventilation 107, 110, 113
watering 98, 102, 107, 109
  not in June to mid-July 98

Violets, varieties and cultivars
(A) botanical names:
Viola *acaulis, foliis ex argenteo variegatis...* 166
  alba 188
  arborea 180, 183
  arborescens 180
  biflora 73, 169, 179
  canadensis 179
  canina 72, 172
  collina 28
  cornuta and varieties 179
  cucullata 72, 75, 76, 103, 179
  cucullata alba 72, 103, 169, 179
  cucullata grandiflora 72, 73, 94, 103, 169, 179
  cucullata striata 72, 103, 169, 179
  elatior 181
  erecta 180
  lusitanica flore caeruleo odoratissima 165
  martia arborea 180
  martia arborescens 181
  obliqua 188
  odorata 150, 172, 180
  odorata leucanthera 28
  odorata purpurea 28
  odorata sulphurea 27, 141, 174
  palmata 27
  pensylvanica 169
  permixta 28
  pubescens 27, 73, 76, 94, 169, 179
  purpurea 180
  sepincola 28
  suavis 157
  subcarnea 28
  sulfurea 174
  vilmoriniana 174
(B) cultivars:
a) single flowers ('Cyclope' and 'Lloyd George' are semi-double; 'Princesse de Sumonte' was thought to be a single Parma violet)
  'Alba Simplex' 175, 178

Index

'Amiral Avellan' 27, 70, 87, 90, 97, 106, 114, 168, 172, 177, 192, 194
'Argentiflora' 84, 178
'Armandine Millet' 11, 19, 27, 62, 90, 100, 124, 165, 175, 177
'Askania' 177
'Baronne de Rothschild' 177, 194
'Bleue de Fontenay' 27, 64, 87, 88
'Brune de Bourg-la-Reine' 11, 27, 61, 90, 97, 165, 175, 177
'Bruneau' 156
'California' 27, 177
'Coeur d'Alsace' 20, 28, 175, 177
'Comtesse Edmond Dutertre' 27, 71, 168, 177, 192, 194
'Cyclope' 177
'Czar' 47-49, 56, 58, 59, 61, 62, 64, 85, 87, 88, 97, 106, 114, 157, 164, 166, 168, 177, 192, 194
'Czar Blanc' (='White Czar')
'Czar bleu' 48, 61, 67
'Dybowski' (='Explorateur Dybowski) 70, 106
'Explorateur Dybowski' 27, 89, 97, 175, 177
'Gloire améliorée' 69, 168
'Gloire de Bourg-la-Reine' 14, 27, 48, 58, 62, 67, 68, 70, 71, 89, 97, 106, 165, 168, 175, 177, 194
'grande Luxonne, la' (='Mme.E.Arène') 48
'grosse bleue' 54
'Helvetia' 28, 175
'l'Inépuisable' 64, 87, 88, 166, 192
'Irish Elegance' 174
'John Radenbury' 177
'La France' 26, 27, 71, 90, 147, 175, 177, 192, 194
'Lianne' 175, 177
'Lilas' 20, 27, 60, 61, 85, 88, 106, 175, 178
'Lloyd George' (='Mrs.David Lloyd George') 177
'Luchon' (='Luxonne') 63
'Luxonne' 14, 26, 48, 49, 56, 63, 69-71, 87, 90, 114, 157, 172, 175, 177, 194
'Mme.E.Arène' 27, 48, 177, 186, 188, 192
'Mme.Laredo' 175, 177
'Mme.Pagès' 27
'Mlle.A.Pagès' 178

'Mlle.Bonnefoy' 175, 177
'Mlle.Garrido' 28
'Mlle.Louise Tricheux' 175, 178
'Mlle.Schwartz' 178
'Mlle.Susanne Lemarquis' 175, 177
'Marietta' 28, 175
'Mignonette' 85, 178
'Millet père' (='Souvenir de Millet père') 58
'Mrs.Pinchurst' 178
'Nana compacta' 27, 175, 178
'Noélie' 178
'Odorata rubra' 62, 85, 87, 90, 97, 127, 178
'Opéra' 175, 177
'Perle Rose' 20, 178
'Princesse Béatrice' 71, 168
'Princesse de Galles' 15, 26, 27, 48, 49, 67-72, 90, 97, 106, 114, 147, 168, 172, 175, 178, 194
'Princesse de Sumonte' ('Principessa di Sumonti') 27, 71, 155, 169, 178
quatre-saisons violets 49, 51, 53, 54, 56, 84, 88, 166
'Quatre-saisons bleue' 64, 87, 89, 178
'Quatre-saisons hâtive' 114, 115, 178
'Quatre-saisons odorante' 47, 49
'Quatre-saisons ordinaire' 97, 106, 115
'Quatre-saisons Semprez' 56, 57, 59, 64, 87, 89, 106, 114, 115, 178
'Ravageot' 54, 160
'Rawson's White' 85, 87, 90, 127, 170, 178
'Reine Augustine' 27, 178
'Reine Victoria' 48, 61, 67, 157, 178
'Rosea delicatissima' 28, 175
'Russian, The' (='Czar') 47, 59, 114, 115, 166
'Sans Éperon' 178
'Sans Pareille' 11, 27, 175
'Sans Prix' 27
'Souvenir de Ma Fille' 20, 28, 175, 177
'Souvenir de Jules Josse' 178
'Souvenir de Millet père' 10, 11, 27, 48, 89, 100, 106, 175
'Subcarnea' 178
'Sulphurea' 141, 178

'Tigrée or' (='Viola à fleurs tigrées or') 66, 87, 90, 167, 170, 178
'Udine' 178
'Velsiana' (='Wellsiana') 178
'Victoria' 87
'Violette de Constantinople' (='Wilson') 47, 157
'Wellsiana' 67, 87, 88, 167, 178
'White Czar' 27, 61, 72, 85, 127, 177, 192, 194
'Wilson' 47, 56, 62, 63, 97, 127, 157
'Wilson Extra' 63
'Wilson, Grosse' 63
b) hardy double violets. This list contains many synonyms and is given merely for reference to the text. 'Blanche de Chevreuse', 'Comte de Chambord' (similar to the description of 'Belle de Châtenay'), 'Rose double' and one double blue are about the only survivors today.
'Belle de Châtenay' ('Grandiflora tricolor') 60, 90, 137, 178
'Blanche double de Chevreuse' 87, 102, 137, 138, 178
'Bleue double' 90, 178
'Double bleue' 81, 87
'double blue' 38, 42, 55, 137, 151
'double pink' 38, 151
'double red' 44
'double rose' 46
'double white' 44, 46, 137
'en arbre' 102, 154, 171, 182
'King of Violets' 178
'La Parisienne' 178
'Louise Baron' 178
'Patrie' 27, 88, 102, 154, 178, 183, 193
'Princesse Irène' 178
'Roi des Violettes' 27
'Rose double' 155, 178
'Rose double de Brunaut' 27
'Rose double de Bruneau' 178
'Tree violet' 42, 44
c) Parma violets
'Brandyana' illustr.185
'Brazza White' (='Comte de Brazza') 64
'Comte de Brazza' 92, 109, 110, 166, 167, 179
'Gloire d'Angoulême' 92, 157, 179
'Lady H.Campbell' 179
'Mme.Millet' 27, 28, 64, 92, 93, 166, 167, 175, 179
'Marguerite de Savoie' 59, 179
'Marie-Louise' 59, 92, 109, 120, 164, 179
'M.J.Astorg' ('Mrs.J.J.Astor') 27, 179
'Neapolitan Violet' 155
'Parme d'Angoulême' 49
'Parme de Toulouse' 28, 46, 60, 92, 157, 165, 179
'Parme de Turquie' 92
'Parme ordinaire' 28, 46, 64, 65, 92, 179
'Parme sans filets' 46, 54, 92, 102, 161, 179
'Président Poincaré' 177, 179
'Queen Mary' 177
'Swanley White' (='Comte de Brazza') 27, 28, 64, 65, 92, 166, 179
'Venice' 166

violet-sellers 51, 52
Violette, Mme.veuve 24
Violettes, Les 175, 185
Violettes à Hyères, Les 168
Virgil 36, 37, 149, 150, 153
Virgil, *Eclogues* 36, 37, 153
Vitruvius 37, 150
Vogt, Armandine Madeleine 160
François Xavier 3, 7, 54
Joseph 160
Marie Armandine 160
Xavier Harmand 160

Ware, T.S. 59, 164
Weston 166
whortleberry 37, 38
wicker baskets, trays 112, 137
Wilson, Edward 156

Yvon, J.B. 59, 164, 169